Current Strategies to Improve the Nutritional and Physical Quality of Baked Goods

Current Strategies to Improve the Nutritional and Physical Quality of Baked Goods

Special Issue Editors

Mario Martinez Martinez
Manuel Gómez Pallarés

MDPI • Basel • Beijing • Wuhan • Barcelona • Belgrade • Manchester • Tokyo • Cluj • Tianjin

Special Issue Editors

Mario Martinez Martinez
University of Guelph
Canada

Manuel Gómez Pallarés
University of Valladolid
Spain

Editorial Office
MDPI
St. Alban-Anlage 66
4052 Basel, Switzerland

This is a reprint of articles from the Special Issue published online in the open access journal *Foods* (ISSN 2304-8158) (available at: https://www.mdpi.com/journal/foods/special_issues/baked_goods_Nutritional_Physical).

For citation purposes, cite each article independently as indicated on the article page online and as indicated below:

LastName, A.A.; LastName, B.B.; LastName, C.C. Article Title. *Journal Name* **Year**, *Article Number*, Page Range.

ISBN 978-3-03928-346-0 (Pbk)
ISBN 978-3-03928-347-7 (PDF)

© 2020 by the authors. Articles in this book are Open Access and distributed under the Creative Commons Attribution (CC BY) license, which allows users to download, copy and build upon published articles, as long as the author and publisher are properly credited, which ensures maximum dissemination and a wider impact of our publications.
The book as a whole is distributed by MDPI under the terms and conditions of the Creative Commons license CC BY-NC-ND.

Contents

About the Special Issue Editors vii

Mario M. Martinez and Manuel Gomez
Current Trends in the Realm of Baking:
When Indulgent Consumers Demand Healthy Sustainable Foods
Reprinted from: *Foods* **2019**, *8*, 518, doi:10.3390/foods8100518 1

Laura Roman and Mario M. Martinez
Structural Basis of Resistant Starch (RS) in Bread: Natural and Commercial Alternatives
Reprinted from: *Foods* **2019**, *8*, 267, doi:10.3390/foods8070267 5

Zhiguang Huang, Letitia Stipkovits, Haotian Zheng, Luca Serventi and Charles S. Brennan
Bovine Milk Fats and Their Replacers in Baked Goods: A Review
Reprinted from: *Foods* **2019**, *8*, 383, doi:10.3390/foods8090383 25

Andrea Bresciani and Alessandra Marti
Using Pulses in Baked Products: Lights, Shadows, and Potential Solutions
Reprinted from: *Foods* **2019**, *8*, 451, doi:10.3390/foods8100451 45

Beatriz de Lamo and Manuel Gómez
Bread Enrichment with Oilseeds. A Review
Reprinted from: *Foods* **2018**, *7*, 191, doi:10.3390/foods7110191 65

Luís M. Cunha, Susana C. Fonseca, Rui C. Lima, José Loureiro, Alexandra S. Pinto, M. Carlota Vaz Patto and Carla Brites
Consumer-Driven Improvement of Maize Bread Formulations with Legume Fortification
Reprinted from: *Foods* **2019**, *8*, 235, doi:10.3390/foods8070235 79

Simona Grasso, Ese Omoarukhe, Xiaokang Wen, Konstantinos Papoutsis and Lisa Methven
The Use of Upcycled Defatted Sunflower Seed Flour as a Functional Ingredient in Biscuits
Reprinted from: *Foods* **2019**, *8*, 305, doi:10.3390/foods8080305 91

Georgiana Gabriela Codină, Ana Maria Istrate, Ioan Gontariu and Silvia Mironeasa
Rheological Properties of Wheat–Flaxseed Composite Flours Assessed by Mixolab and Their Relation to Quality Features
Reprinted from: *Foods* **2019**, *8*, 333, doi:10.3390/foods8080333 103

Thomas Mellette, Kathryn Yerxa, Mona Therrien and Mary Ellen Camire
Whole Grain Muffin Acceptance by Young Adults
Reprinted from: *Foods* **2018**, *7*, 91, doi:10.3390/foods7060091 119

Mayara Belorio, Marta Sahagún and Manuel Gómez
Influence of Flour Particle Size Distribution on the Quality of Maize Gluten-Free Cookies
Reprinted from: *Foods* **2019**, *8*, 83, doi:10.3390/foods8020083 133

Catrin Tyl, Radhika Bharathi, Tonya Schoenfuss and George Amponsah Annor
Tempering Improves Flour Properties of Refined Intermediate Wheatgrass
(*Thinopyrum intermedium*)
Reprinted from: *Foods* **2019**, *8*, 337, doi:10.3390/foods8080337 143

About the Special Issue Editors

Mario Martinez Martinez (MSc, PhD) finished his PhD in Food Chemistry in 2016 and, shortly after, joined the Whistler Center for Carbohydrate Research, Department of Food Science (Purdue University, IN, USA) as a Postdoctoral Research Fellow. Dr. Martinez started his independent research career as Tenure-Track Assistant Professor at the University of Guelph (ON, Canada) in August 2017. His research focuses on fundamental research to practical applications of edible plant tissues and resorts to physicochemical, biological, and engineering concepts to extend the use of plant-based ingredients regarding functionality and health. He is an emerging expert in the characterization of carbohydrates and their modification through process intensifying technologies, such as high shear extrusion processing. He has published more than 60 peer-reviewed scientific papers on the structure–function of carbohydrate polymers and the interactions of carbohydrate-rich matrices with phenolic compounds. His strengths are in the expansion of such mechanistic techniques across multiple disciplines and on complex biological matrices.

Manuel Gómez Pallarés (MSc, PhD) joined the Food Technology Area (College of Agricultural Engineering, Palencia, Spain) at the University of Valladolid in 1994. During his first few years as an independent researcher, Dr. Gomez created the Cereal Science Laboratory located in Palencia, Spain. His research initially approached the improvement of wheat flours for bread-making. Subsequently, his research program focused on the nutritional improvement of different baked goods and the development of foods adapted to the special needs of certain groups, including the improvement of gluten-free goods. Currently, Dr. Gomez works in the manufacturing of novel flours with improved functionality and the recycling of food byproducts with the objective of reducing food waste. Dr. Gomez has published more than 130 peer-reviewed scientific papers, authored 5 patents, and has performed multiple works for milling and bakery companies.

Editorial

Current Trends in the Realm of Baking: When Indulgent Consumers Demand Healthy Sustainable Foods

Mario M. Martinez [1,*] and Manuel Gomez [2,*]

1. School of Engineering, University of Guelph, Guelph, ON N1G 2W1, Canada
2. Food Technology Area, College of Agricultural Engineering, University of Valladolid, 34004 Palencia, Spain
* Correspondence: mario.martinez@uoguelph.ca (M.M.M.); pallares@iaf.uva.es (M.G.); Tel.: +1-519-824-4120 (ext. 58677) (M.M.M.); +34 979-10-8495 (M.G.)

Received: 14 October 2019; Accepted: 17 October 2019; Published: 21 October 2019

The term "baked goods" encompasses multiple food products made from flour (typically wheat flour). Among them, bread has stood as a foundation in different cultures by providing energy, mostly from its starch fraction, while being low in fats and sugars. Nevertheless, breadcrumbs are categorized as having a high amount of rapidly digestible starch that has been associated with poor health outcomes, including type 2 diabetes, obesity, and cardiovascular disease, as well as other metabolic-related health problems. In this regard, the enrichment of bread with resistant starch (RS) ingredients is gaining prominence and can be definitively positioned as an impactful strategy to improve human health through diet. In this Special Issue, structural factors for the resistance to digestion and hydrothermal processing of clean label RS ingredients are reviewed by Roman et al. [1], who expanded the definition of each RS subtype to account for recently reported novel and natural non-digestible structures. The term baked goods also include cakes and cookies, which are rich in fats and sugars but represent an excellent choice for indulgent consumption. While bread may be an excellent food carrier of added nutritional and extranutritional compounds, such as proteins, dietary fibers and bioactive phytochemicals, the effort to improve the nutritional properties of cakes and cookies has focused on the elimination or reduction of fats and sugars associated with poor health outcomes. As an example, milk fats have typically been used in cake- and cookie-making, and their high content in calories and saturated fatty acids has encouraged food researchers and technologists to develop fat mimetics, as discussed in this Special Issue in the review by Huang et al. [2].

Many research groups have focused on the enrichment of baked goods with other plant-based ingredients of high nutritional value. Legume flours possess a high content of proteins with an amino acid profile complementary to that of cereals. As a result, the enrichment of breads with these flours has received significant attention over the last years, as revised in this Special Issue by Bresciani and Martí [3]. However, the incorporation of legume flours into baked goods usually results in lower organoleptic quality and the recipe must be re-adjusted to minimize these detrimental effects, as reported by Cunha et al. [4]. The use of ingredients from oil seeds is also becoming paramount in many recipes over the last years because they possess higher protein content than cereals and are rich in fiber, omega-6 and omega-3 essential fatty acids, and natural antioxidant compounds, including tocopherol, beta-carotene chlorogenic acid, caffeic acid and flavonoids. As discussed in the review written by De Lamo and Gómez [5], oil seeds can be added directly as whole seeds or as milled flour. In this Special Issue, Grasso et al. [6] considered the enrichment of cookies with defatted sunflower seed flour and Codina et al. [7] investigated the use of flaxseed flour in bread-making. As observed in these works, the nutritional improvement of baked goods derived from the use of the aforementioned nutrient-dense ingredients almost always worsens their physical quality. This may result in a critical loss of consumers' acceptance and, therefore, the unfeasible translation of nutrient-dense ingredient incorporation to the commercial reality. This aspect is approached by Mellette et al. [8] using cakes made with whole flour. In this regard, Belorio et al. [9] found that optimization of the physical properties of a flour, specifically in terms of particle size, dramatically impacted the physical qualities of their baked good: cookies. In their work, the authors encourage ingredient technologists to optimize clean and simple technologies, such as milling mechanical fractionation, to produce clean label flours with optimum physical properties and successful commercial applications.

Baked goods are also characterized as having a low protein content, although their high consumption makes them account for a significant fraction of the total recommended protein uptake. Nonetheless, protein scores in cereals, which are commonly the main ingredients in baked goods, are usually low due to a suboptimal amino acid profile and low protein digestibility. Interestingly, the overall protein digestibility is not only dependent on the protein source, but also the food processing methodology. The review written by Joye [10] provides an in-depth evaluation of protein digestibility as affected by the typical unit operations carried out during the manufacture of baked goods.

Last but not least, this Special Issue considers the consumers' increased awareness of the environment and sustainable food systems. In this regard, novel processing and breeding technologies have been reported as key contributors to reduced food waste and loss. As an example, the use of perennial grains has been reported to result in more efficient use of water, fertilizers, and soil nutrients, although their incorporation into foods is only possible if the quality of their resultant flours matches the expectations of both manufacturers and consumers. In this Special Issue, the impact of milling and tempering on the perennial grain intermediate wheatgrass was studied by Tyl et al. [11].

The works included in this Special Issue highlight the importance of holistically considering the nutritional improvement of baked goods by using sustainable plant-based ingredients and the optimization of the physical properties of such ingredients to result in successful commercial applications. However, scientists and technologists within the realm of baking should invest in translational research that provides a detailed understanding of food and food ingredient nano- and micro-structures, as well as the impact of processing and the development of successful recipes.

Author Contributions: The authors contributed equally.

Conflicts of Interest: The authors declare no conflict of interest.

References

1. Roman, L.; Martinez, M.M. Structural basis of resistant starch (RS) in bread: Natural and commercial alternatives. *Foods* **2019**, *8*, 267. [CrossRef] [PubMed]
2. Huang, Z.; Stipkovits, L.; Zheng, H.; Serventi, L.; Brennan, C.S. Bovine milk fats and their replacers in baked goods: A review. *Foods* **2019**, *8*, 383. [CrossRef] [PubMed]
3. Bresciani, A.; Marti, A. Using pulses in baked products: Lights, shadows, and potential solutions. *Foods* **2019**, *8*, 451. [CrossRef] [PubMed]
4. Cunha, L.M.; Fonseca, S.C.; Lima, R.C.; Loureiro, J.; Pinto, A.S.; Vaz Patto, M.C.; Brites, C. Consumer-driven improvement of maize bread formulations with legume fortification. *Foods* **2019**, *8*, 235. [CrossRef] [PubMed]
5. De Lamo, B.; Gómez, M. Bread enrichment with oilseeds. A review. *Foods* **2018**, *7*, 191. [CrossRef] [PubMed]
6. Grasso, S.; Omoarukhe, E.; Wen, X.; Papoutsis, K.; Methven, L. The use of upcycled defatted sunflower seed flour as a functional ingredient in biscuits. *Foods* **2019**, *8*, 305. [CrossRef] [PubMed]
7. Codină, G.G.; Istrate, A.M.; Gontariu, I.; Mironeasa, S. Rheological properties of wheat–flaxseed composite flours assessed by mixolab and their relation to quality features. *Foods* **2019**, *8*, 333. [CrossRef] [PubMed]
8. Mellette, T.; Yerxa, K.; Therrien, M.; Camire, M.E. Whole grain muffin acceptance by young adults. *Foods* **2018**, *7*, 91. [CrossRef] [PubMed]
9. Belorio, M.; Sahagún, M.; Gómez, M. Influence of flour particle size distribution on the quality of maize gluten-free cookies. *Foods* **2019**, *8*, 83. [CrossRef] [PubMed]

10. Joye, I. Protein digestibility of cereal products. *Foods* **2019**, *8*, 199. [CrossRef] [PubMed]
11. Tyl, C.; Bharathi, R.; Schoenfuss, T.; Annor, G.A. Tempering improves flour properties of refined intermediate wheatgrass (*Thinopyrum intermedium*). *Foods* **2019**, *8*, 337. [CrossRef] [PubMed]

© 2019 by the authors. Licensee MDPI, Basel, Switzerland. This article is an open access article distributed under the terms and conditions of the Creative Commons Attribution (CC BY) license (http://creativecommons.org/licenses/by/4.0/).

Review

Structural Basis of Resistant Starch (RS) in Bread: Natural and Commercial Alternatives

Laura Roman and Mario M. Martinez *

School of Engineering, University of Guelph, Guelph, ON N1G 2W1, Canada
* Correspondence: mario.martinez@uoguelph.ca; Tel.: +1-519-824-4120 (ext. 58677)

Received: 1 July 2019; Accepted: 16 July 2019; Published: 19 July 2019

Abstract: Bread is categorized as having a high amount of rapidly digested starch that may result in a rapid increase in postprandial blood glucose and, therefore, poor health outcomes. This is mostly the result of the complete gelatinization that starch undergoes during baking. The inclusion of resistant starch (RS) ingredients in bread formulas is gaining prominence, especially with the current positive health outcomes attributed to RS and the apparition of novel RS ingredients in the market. However, many RS ingredients contain RS structures that do not resist baking and, therefore, are not suitable to result in a meaningful RS increase in the final product. In this review, the structural factors for the resistance to digestion and hydrothermal processing of RS ingredients are reviewed, and the definition of each RS subtype is expanded to account for novel non-digestible structures recently reported. Moreover, the current in vitro digestion methods used to measure RS content are critically discussed with a view of highlighting the importance of having a harmonized method to determine the optimum RS type and inclusion levels for bread-making.

Keywords: high-amylose; digestion; bakery; retrogradation; glycemic response; amylose; amylopectin; α-amylase

---

1. The Importance of Bread in the Human Diet

Carbohydrates are the most important source of dietary energy for humans (45–70% of total energy intake) [1], with starch being the main structure-building macro-constituent in many foods, including bread, pastry, breakfast cereals, rice, pasta, and snacks. White bread, with an average consumption of about 170 g per day per person in 10 European countries, contributes to the highest proportion of carbohydrates to the daily dietary intake [2]. Despite current findings showing dose-response relation between consumption of whole grains and the risk of non-communicable diseases [3], white wheat bread remains consumers' first choice mainly owing to its sensory attributes [4]. This event remarkably highlights the technological challenge of the incorporation of dietary fibers to make palatable breads acceptable by consumers, that is, the type and amount of dietary fiber ingredients must be meticulously selected based on their impact on bread quality [5].

Besides lacking the nutritional components from the whole grain fraction, white bread is categorized as having a high amount of rapidly digestible starch. This is the result of starch gelatinization produced as a consequence of the high temperatures that the dough reaches during baking (≥70 °C) at relatively high-water content (≥35%) [6,7]. In fact, a complete starch gelatinization in white bread crumb almost always occurs [6,8,9]. In this regard, consumption of white breads, which results in a rapid increase of the postprandial blood glucose, is associated with poor health outcomes including type 2 diabetes, obesity, cardiovascular disease, as well as other metabolic-related health problems [10–12].

In view of the large consumption of daily white bread and the health benefits associated with higher dietary fiber consumption [13], the enrichment of bread crumbs with resistant starch (RS)

ingredients is gaining prominence (Figure 1) and can definitively be positioned as an impactful strategy to improve human health through the diet. A literature search in the topic also revealed significantly more studies of RS in breads than in cakes, muffins, and cookies. Because the RS property can change during baking, this review will cover the structural factors responsible for the RS digestion property and the thermal stability of RS ingredients to manufacture breads with meaningful health outcomes. In this review, the structural basis for the RS property of RS in breads will be revised based on recent pivotal studies. Furthermore, the definition of RS will be discussed, addressing holistically and briefly the current analytical methods for quantifying the RS content of foods and the current regulations in terms of food labeling and health claims. We expect that this review provides a brief overlook of the currently commercially available RS ingredients, with special focus on those that support clean and natural labels (i.e., RS4 will not be discussed).

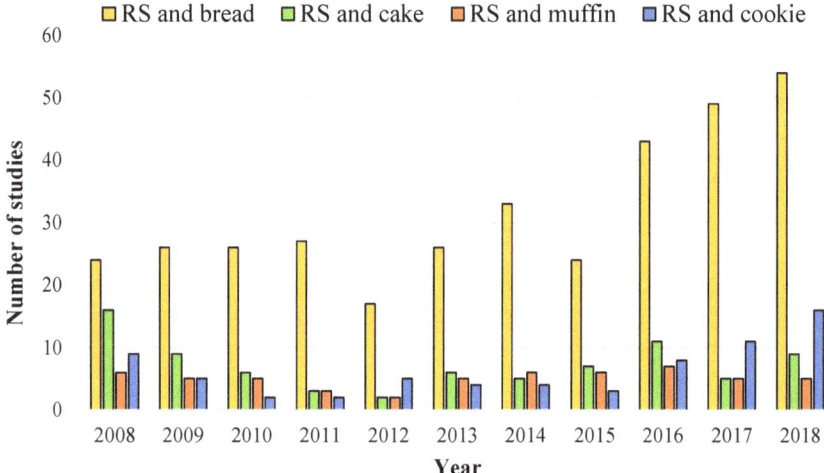

Figure 1. Literature search of the last 10 years on the topics: "resistant starch (RS) and bread"; "resistant starch and cake"; "resistant starch and muffin"; and "resistant starch and cookie". Data collected from all databases from the Web of Science on 28 June 2019.

2. RS Definition and Analytical Methods

Resistant starch (RS) is defined as the starch portion that escapes digestion by human enzymes in the upper part of the gastrointestinal tract, entering the large intestine where it can be partially or fully fermented by colonic microflora. The main health outcomes of RS consumption can be categorized mainly based on a modulation on the glycemic response, body weight control, and bowel health. However, this review is not intended, by any means, to provide deep insights into the complex effects of RS consumption on specific metabolic responses and health benefits, which has been previously revised elsewhere [5,14–27].

According to its definition, RS should be predicted by physiological (in vivo) techniques [28], such as the human ileostomy model, where ileal digesta from adults with permanent ileostomies is analyzed for its starch content and compared with the total amount of starch ingested during the study period [16]. However, in vivo methods are remarkably slow and tedious, and require a considerable investment in specialized resources and expertise. Added to that, the rate and extent of starch digestion depends on both extrinsic (e.g., chewing, hormone responses, enzyme activity, passage rate, individual health) and intrinsic (food structure) factors, with the former providing a high variability included in in vivo experiments. On the other hand, the variability from extrinsic factors is excluded in in vitro methods, enabling information for understanding the mechanism of food structural changes during the digestion time course [16].

Many in vitro assays for RS determination are variations on Berry's [29] modification of Englyst's original method [30]. Starchy products "as eaten" are subjected to gastric (protease) and luminal (pancreatic α-amylase) digestions under fixed physiological conditions of temperature, pH, viscosity, and rate of mechanical mixing similar to those in the gastrointestinal tract. RS is determined by difference between total and digestible starch [31], with validated in vivo results using the ileostomy model [32]. Digestion products are obtained at 20 and 120 min of incubation with α-amylase and further converted to glucose for colorimetric [31] or chromatographic quantification [33]. In the Englyst test, rapid digestible starch (RDS) is the starch digested fraction within the initial 20 min digestion, slowly digestible starch (SDS) is the digested fraction between 20 and 120 min, and RS is the remaining portion after 120 min.

In 2002, McCleary and Monaghan [34] also developed a wide spread method to determine RS, which was validated by both the Association of Official Analytical Chemists [35] (AOAC Method 2002.02) and the American Association of Cereal Chemists [36] (AACC Method 32-40.01). In this case, starchy foods are simultaneously incubated with pancreatic α-amylase and amyloglucosidase for 16 h (vs. 3 h in the Englyst test) in order to hydrolyze and solubilize all the digestible starch. The non-digested starch, the RS fraction, is recovered after several washes and centrifugation steps, and the RS pellet is dissolved with potassium hydroxide prior its hydrolysis to glucose and colorimetric determination. Several other methods were also proposed for analytical determination of RS [37–41].

RS can also be measured following the procedures used for dietary fiber determination. However, attention should be paid on the methodology used because some RS sources can be underestimated. Thus, the Prosky [42] and Lee [43] methods, as well as AOAC official methods 985.29 (AACC 32-05.01) and 991.43 (AACC 32-07.01), respectively, do not quantitatively measure all the RS. Because of the initial heating step at above 90 °C, thermally unstable RS fractions, such as RS2 from banana or potato, are partially degraded. To alleviate this problem, an integrated procedure for the measurement of total dietary fiber (AOAC Methods 2009.01/2011.25; AACC Methods 32-45.01/32-50.01), which fully includes RS (in the same way as in AOAC 2002.02) and other non-digestible oligosaccharides [44,45], was proposed. Therefore, the combination of AOAC 2009.01 and 2002.02 methods could provide quantitative determination of total dietary fiber (including all the RS fractions) and RS, respectively. However, because of the simplicity of AOAC 2002.02, this procedure is recommended if only RS is the dietary fiber of interest.

RS is usually categorized following the RS classification given by Englyst, Kingman, & Cummings [31]; Eerlingen & Delcour [46]; and Brown et al. [47] based on the structural features conferring its resistance. In this way, RS is usually listed into five categories, as follows. RS1: physically entrapped, non-accessible starch in a non-digestible matrix; RS2: native granular resistant starch (B- or C-polymorph); RS3: retrograded starch; RS4: chemically modified resistant starch; and RS5: single amylose helix complexed with lipids. In Table 1, the structural features conferring the RS property within each category (reported to date) are listed and categorized based on the RS classification given by Englyst, Kingman, & Cummings [31]; Eerlingen & Delcour [46]; and Brown et al. [47]. Although this traditional categorization is the most used to date, it is noteworthy that it assumes RS to be a thermodynamically defined structural form (physical entities) and discards its potential kinetic nature. If RS was simply thermodynamically defined, only highly chemically-modified starches (RS4) would be completely resistant to enzyme hydrolysis. This is a critical point in bread-making, as flour/starch fabrication and baking will strongly alter the RS type and content [6,48–50]. As an example, baking will generally destroy RS1 and RS2, but may form RS3 and RS5, generally resulting in breads containing RS < 2.5% (dry matter) [40]. In this section, the structural types of RS listed in Table 1 will be briefly described and linked to their effects on bread physical and nutritional quality. Special attention will be put on commercially available RS2 and RS3 clean ingredients (see Section 4 and Table 2). Resistant maltodextrins, soluble chemically modified-dextrins derived from starch and included in the definition of RS, are also commercially available. However, this review will only focus on RS excluding starch degradation products that may also be resistant to digestion by pancreatic α-amylase.

Table 1. Structural features conferring the resistant digestion property within each clean-label resistant starch (RS) category.

Classification	Structural Features Conferring the RS Property within Each Category	Detrimental Steps That May Decrease RS Content during Bread-Making	Assisting Steps That May Increase RS Content during Bread-Making
RS1	Intact plant tissues	Milling, sieving, baking	-
	Highly dense food matrices	-	Baking and cooling
	Confined starch within a continuous layer of certain proteins	-	Baking of starch materials containing specific layer forming proteins
RS2	Starch granules with an outer high-density shell structure	Baking (of note that high amylose RS2 is more heat-resistant)	-
RS3	Retrograded amylose	-	Baking and cooling
	High-density processed amylose	-	Extrusion of high amylose starch ingredients
	Retrograded amylopectin	Baking	Baking and cooling
RS4	Chemically substituted starches	-	-
	Chemically cross-linked starches	-	-
	[a] Resistant maltodextrins	-	-
RS5	Amorphous amylose-lipid complexes (form I)	-	Baking and cooling
	Crystalline amylose-lipid complexes (form II)	-	Baking and cooling

[a] Resistant maltodextrins can be defined as chemically-modified dextrins instead of chemically-modified starch. In that case, they should be excluded from this list.

3. Natural RS Ingredients in Bread-Making and Structural Basis of Their Resistant Digestion

3.1. Physical Barriers Comprising Plant Cell Walls and/or the Food Matrix (RS1)

The resistant digestion property of RS can be the result of its confinement within the intact plant cell (surrounded by the plant cell wall) and/or the food matrix. Overall, the role of cell walls in limiting starch digestion is based on three mechanisms [51–57]: (1) the difficulty for amylase to permeate through the cell wall; (2) the limitation of starch gelatinization during cooking; and (3) the binding of α-amylase by cellulose and other cell wall components. Whole or partly milled grains or seeds with intact cell walls are clear examples of physically confined starch within cell walls. Milling should be performed carefully to avoid the loss of RS1, as the tissue matrix (cell wall and protein network) could be damaged [57,58]. The effects can be minimized with coarse milling or selection of large particles after mechanical fractionation [57,59]. Nonetheless, large particles are not always suitable and the selection of plant materials with thicker and less permeable cell walls, such as legume flours [52,54] or cereal flours from hard endosperm [57], could increase the content of starch that escapes digestion entirely, even after cooking.

The presence of whole or partly milled grains and seeds has been reported to decrease the glycemic index of breads [60,61]. However, the use of intact kernels (or broken kernels) will always impact significantly the bread physical and sensory properties. Therefore, food technologists should bear in mind that white bread is the most consumed bread type nowadays [4]. There is little doubt about the health benefits associated with a higher consumption of whole grains [3]. However, to what extent can the particle size of intact grains be reduced to result in breads with lower starch digestion (glycemic response)? Interestingly, Edwards et al. [55] demonstrated that fully cooked and gelatinized porridges, made with 2 mm wheat flour particles, resulted in significantly lower blood glucose, insulin, C-peptide, and glucose-dependent insulinotropic polypeptide concentrations than porridges made with <0.2 mm particles. In fact, they showed that the structural integrity of coarse wheat particles was retained during gastroileal transit using a randomized crossover trial in nine healthy ileostomy participants. However, flours for bread-making are usually smaller than 250 µm and complimentary studies should be performed with smaller variations in particle size. Martinez, Calvino, Rosell, & Gomez [62] observed that among <250 µm flour particles, a differential of 100 µm (coarser) can result in a lower rate and extent of starch digestion, even after full gelatinization through high-shear extrusion. Nevertheless, their effect after incorporation into breads has received little attention. Only de la Hera et al. [63] observed that breads made with coarser rice flour (132–200 µm) presented higher RS than those made with fine flours (<132 µm). On the other hand, Protonotariou, Mandala, and Rosell [64] did not observe differences in the amount of RS between breads made with whole wheat flours with particle size ranging from 57 to 120 µm. Remarkedly, these two studies included the RS values of bread samples after freeze-drying and milling. Even if freeze-dried crumb samples were corrected for moisture and sieved to discard particle size effects, this approach for sample preparation still disregards potential changes in the permeability of the intact plant cell and/or the food matrix. In any case, differences in RS were small and human intervention studies should be performed to confirm or discard the use of coarse flours feasible for bread-making for better postprandial metabolism. Added to that, it should be noted that the amount of ungelatinized starch is dramatically higher in bread crust than in bread crumb [6], and hence the effects of varying particle size could be completely different between crumbs and crusts. In this sense, de la Hera et al. [63] and Protonotariou, Mandala, & Rosell [64] investigated the RS content in bread slices containing the crust portion, so the question of whether particle size differences in the range of 100 µm affect RS in bread crumb, the major fraction of a bread slice, remains unclear.

Besides plant cell walls, storage proteins from certain plants, such as those from wheat (glutenin and gliadin), maize (zein), and sorghum (kafirin), have the ability to form disulphide bonds that result in a continuous layer around starch granules upon cooking, and in a slowdown of starch digestion [65,66]. In any case, the effect of network-forming proteins on the resulting RS (or glycemic response) after baking has received little attention. Only Berti et al. [67] and Jenkins et al. [68] showed lower postprandial glucose levels of gluten-containing breads compared with gluten-free breads, which was attributed to the presence of a protein network encapsulating the starch. Jenkins et al. [68] also proved that the addition of gluten to gluten-free breads did not reduce the glycemic response, suggesting that the protective effect of the protein present in the wheat is the result of the natural junctions between protein and starch, and is lost once the protein–starch network is disrupted. On the other hand, zein and kafirin, presumably owing to their relative hydrophobicity and disulphide bond cross-linking [69], are isolated in protein bodies in the endosperm cells of the mature grain [70]. The localization of storage proteins in protein bodies, unlike what occurs in wheat, prevents the formation of a continuous matrix around the starch granules within the cells. For zein and kafirin to be functional in doughs, the protein bodies must be disrupted during dough mixing and the proteins freed. However, disruption of the protein bodies has only been observed to occur during high shear extrusion [71] or roller flaking [72].

3.2. Granular Surface Properties (Granular Resistant Starch, RS2)

Starch usually gelatinizes in the range of 54 to 76 °C at ≥20% water [73]. Therefore, considering that, even for those breads made with the lowest possible hydration level (refined dough bread, also known as candeal bread), the moisture content in the crumb is above 35% throughout baking (where a temperature above 70 °C is reached [7]), an extensive (mostly complete) starch gelatinization (Figure 2) is expected to occur [6]. On the contrary, the fast evaporation of water from the crust owing to its high surface temperature impairs the full gelatinization of the starch [6]. In this way, it is possible to find from 56% to 70% (or even higher) of the starch in the crust ungelatinized (Figure 2), depending on the type of bread [6,9]. Restriction of swelling and gelatinization can also be achieved by the interplay of starch with other ingredients in the formula, including lipids, protein, fibers, and sugars [74]. In any case, the presence of starch granules inherently resistant to digestion (RS2) could increase the final content of RS in breads coming from their crust portion [9].

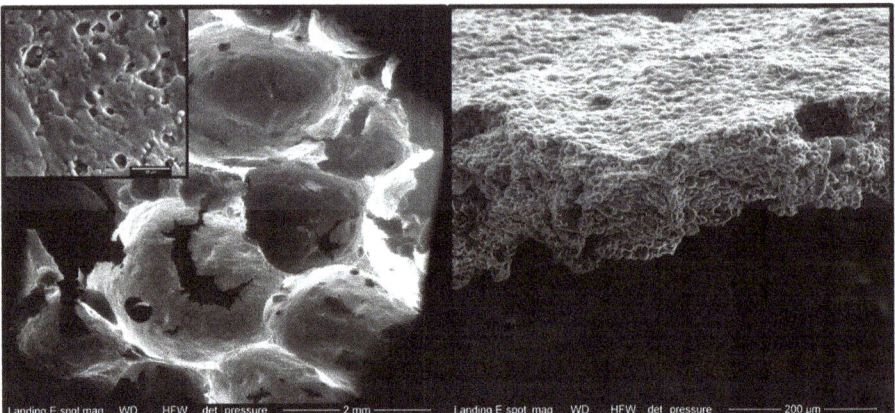

Figure 2. Micrographs of crumb (left) and crust (right) sections of breads containing 20% of RS2 banana starch. Detailed magnification (20 µm) denotes the presence of some granules in the gelatinized crumb.

RS2 has been found in ungelatinized tubers, particularly in potato, as well as in starchy fruits, such as green banana, both in vitro [31] and in vivo [32,75]. High-amylose starch is also a source of RS2. High-amylose starch, which is found mainly in maize, is obtained by mutation of the amylose-extender (ae) gene and the gene encoding starch branching-enzyme I [15]. Thus, this starch presents longer branch chains of intermediate material and higher amylose content [76]. RS2 starches are present in starch granules containing the B-type crystalline allomorph. Although differences in the crystalline structure help explain the higher resistance to amylolytic enzymes of potato, high amylose, and banana starches, crystallinity itself does not fully explain the resistance of these starches. At a superior level of starch structure, A-polymorphic starches are reported to have pores (0.1 to 0.3 µm diameter) and channels (0.007 to 0.1 µm diameter) through which α-amylase (around 3 nm radius) could diffuse [77]. On the contrary, larger "blocklets" at the periphery of B-type polymorphic starch granules result in the absence of pores and channels [78], which could significantly limit the enzyme digestion, and possibly be the primary determinants for the RS property [25,79].

In general, the addition of RS2 ingredients may result only in a moderate increase of RS in the final bread, as gelatinization will destroy their semi-crystalline granular structure. This moderate increase will be the result of remaining ungelatinized granules in the crust, which represents a significant, but lower portion of the bread slice. As an example, Roman, Gomez, et al. [9] observed an RS increase from 0.26 to 5.66% in the crust with the replacement of the main starchy ingredient by native banana starch, but no significant RS increase was observed in the crumb portion. On the contrary, native high amylose is the only RS2 source that resists gelatinization, making this starch more suitable for hydrothermally-processed foods. In fact, complete gelatinization of these mutant starches is only achieved at temperatures higher than 120 °C [5,80,81]. In addition, once gelatinized, high amylose starches can form high amounts of RS3 [82]. Thus, several types of resistant starch, namely, RS2, RS3, and RS5, can coexist in the final bread.

3.3. Dispersed Starch Molecules Forming Resistant Starch upon Cooling and Storage (RS3)

After gelatinization, which results from baking, dispersed starch molecules begin to re-associate upon cooling, forming tightly packed structures stabilized by hydrogen bonding that are more resistant to digestion [83]. The resistance of retrograded amylose to α-amylase digestion was demonstrated both in vitro and in vivo long ago [84], which was termed as RS3. The amount of RS3 produced from retrograded amylose is dependent on the amylose ratio and its chain length [18,85]. Similar to RS2, the enzyme resistance of RS3 has been associated with the formation of a highly thermostable B-type crystalline structure. Thus, the increased crystallinity is expected to result in fewer available α-glucan chains to which α-amylase can bind and thus reduce the susceptibility of retrograded starch to digestion [82]. Nonetheless, crystallinity itself does not fully explain the resistance of RS3, as previously mentioned with RS2. Amorphous material in enzyme-resistant fractions has been found, confirming that the resistance is not simply based on a specific crystalline structure that is fully undigested [86]. Cairns et al. [87] and Gidley et al. [88] suggested that the resistance to digestion is also the result of other double helices not involved in crystals. More recently, extrusion processing of high amylose starch was shown to result in non-crystalline dense packing of amylose chains upon cooling, which exhibited significantly higher RS content than the cooked counterpart [82,89,90]. Furthermore, the content of RS in extruded high amylose starch was similar to that in a granular native state [82]. We believe that the increase in amorphous RS during extrusion could be the result of the molecular fragmentation of amylose and amylopectin chains during extrusion, which could improve molecular mobility and amorphous molecular packing at submicron length scale. In fact, recently, evidence of shear-induced amylose scission during extrusion has been reported [91].

In contrast to amylose, the branched structure of amylopectin is less prone to retrograde, needing a longer time for the formation of double helical structures [91]. Retrograded amylopectin has been linked to the formation of slowly digestible starch (SDS), and hence to a reduction in the rate of starch digestion [92,93]. Starch with a slow digestion rate has been proposed to partially pass to the large intestine as RS, where it functions as a source to bacterial fermentation [84]. In this way, although RS3 has been generally attributed to the formation of resistant crystalline structures from amylose double helices, some old and recent evidence suggests that retrograded amylopectin should be included as another form of RS3 [84]. In fact, Englyst and Macfarlane [94] already proposed a further classification of RS3 into two subcategories, that is, RS3a and RS3b, comprising retrograded amylopectin and amylose, respectively. In terms of amylopectin, slowly digestible starch structures involving amylopectin have been attributed to the following: (1) high proportions of long chains [93,95]; (2) chains with longer average length [9,92,93]; and/or (3) lower molecular sizes through processes such as acid-hydrolysis or high shear cooking extrusion [9,91,93]. In contrast to RS3 from amylose retrogradation, which is thermally stable (melting of amylose-amylose double helices occurring at ~150–160 °C), double helices or aggregates of double helices involving amylopectin melt at significantly lower temperature (~55 °C) [9,96,97] and, therefore, attention should be paid when using in breads that will be re-heated.

3.4. Introduction of Chemical Structures (Chemically Modified Resistant Starch, RS4)

Starch resistance can also be created by the inclusion of chemical structures along starch chains. The resistance to digestion of chemically modified resistant starch (RS4) is dependent on the type and extent of the chemical modification, mostly consisting of dextrinization, etherification, esterification, oxidation, and/or cross-linking [58].

The mechanisms responsible for the enzymatic resistance of RS4 have been revised elsewhere [98]. It is originated principally by two different mechanisms: (1) the introduction of bulky functional groups (e.g., oxidation, etherification, or esterification with hydroxypropyl, acetyl, and octenyl succinic anhydride groups, among others) and (2) starch cross-linking (typically with phosphate groups). In the former category, large and bulky side functional groups are added by substitution along the α-1,4 D-glucan chains to hinder the enzymatic attack, which also makes adjacent glycosidic bonds inaccessible to the enzymes. As for the latter, the presence of cross-linked starch chains (reaction with two or more hydroxyl groups) inhibits granular swelling, preserves granular integrity (preventing enzyme access), and creates steric hindrance, making amylase unable to properly bind to starch. Furthermore, some of the abovementioned chemical modifications can bring about RS ingredients with up to 68–79% RS. Nonetheless, these chemical methods are characterized by long reaction times (up to 24 h) and environmental concerns (use of excess reagents that need to be properly removed and disposed of). Therefore, this type of modification seems less appropriate nowadays in view of the current health and wellness megatrends, which are orientated to clean and natural (free of chemicals) labels.

RS4 from different starch sources are a widely commercialized RS ingredient, although little information exists from the manufacturer about the nature and level of these modifications. Multinational companies providing RS4 from potato, tapioca, wheat, and/or high-amylose maize include Ingredion [99], Roquette [100], MGP Ingredients [101], and Cargill [102]. Ingredion provides RS4 from high-amylose maize starch and potato starch known as Versafibe 2470 and 1490, respectively. Roquette offers a line of modified starches under the name "CLEARGUM" comprising acetylated, diphosphate, and octenyl succinic anhydride (OSA) starches. MGP ingredients offers a phosphorylated cross-linked starch under Fibersym brand name. Cargill also offers a range of stabilized RS4 starches (C☆PolarTex, C☆StabiTex, C☆Tex), subjected to different chemical modifications (hydroxypropylated, acetylated, phosphorylated starch, and so on). Several research works have focused on the influence of these chemically modified RS starches in RS content, glycemic index, and quality of breads [103–111]. Chemically modified starches preserve their RS property during conventional food hydrothermal processing and, therefore, can significantly increase the RS content in bread. Nonetheless, based on consumers' demands for clean labeled products, these investigations will not be discussed in the present review.

3.5. Lipid Complexed Resistant Starch (RS5)

Amylose can form inclusion complexes with lipids, and these complexes have been shown to be more resistant to digestion [112]. Amylose–lipid complexes naturally exist in some starch sources (principally high amylose starches) [113]. Nonetheless, they can also be formed upon hydrothermal treatments, such as baking, in the presence of exogenous or endogenous lipids (monoglycerides, fatty acids, lysophospholipids, and surfactants) [114]. The stability and resistance to digestion is also dependent on the type of lipid (i.e., carbon unit length and unsaturation) complexed [114–116].

Two forms of complexes can be distinguished depending on their thermostability: Type I amylose–lipid complexes that melt at about 95–105 °C (less ordered structures), and Type II (V-type crystalline structures) melting at about 110–120 °C [117,118].

Although there are no commercially available sources of RS5 in the market, amylose–lipid complexes can also reform upon baking, provided there are lipids in the formula. In this regard, most gluten-free breads, which are mostly made with maize starch and rice flour, incorporate some source of lipid/fat in the formula, to enhance the crumb softness and juiciness, as it tends to be excessively dry [119]. Meanwhile, wheat flour lipids represent 2.0% to 2.5% of the flour and exogenous lipids are often added to reduce hardness or staling [120]. Therefore, the presence of a certain amount of RS5 in breads is expected.

It is worth noting the ~50% reduction of postprandial blood glucose and insulin levels of breads containing 60% (flour basis) of a developed RS5 containing ingredient compared with the control white bread [121]. These authors produced an ingredient containing both RS3 and RS5 by debranching high amylose VII maize starch with isoamylase followed by complexation with palmitic acid. Interestingly, they showed that the debranching treatment increased the amount of linear chains, which could either retrograde or form complex with lipids more effectively (RS: 52.7%) than the native high-amylose starch molecules (RS: 35.4%) upon baking.

4. Effects of RS in Physicochemical Characteristics of the Breads

Although bread-making varies widely around the world, the four basic ingredients are flour/starch (normally from cereals and tubers), water, yeast, and salt. Processing conditions include kneading, proofing, and baking. Inclusion of RS ingredients in the formula is usually given by replacement levels of the starchy material by the RS ingredient. Most investigations approached the RS enrichment of breads using commercial RS2 and RS3 ingredients, but only some studies included the RS content in the final product. This is critical as baking will critically alter the type and amount of RS. For this reason, in Table 2, only those studies in which RS was assessed in the final product were included.

High amylose starches, usually from maize, in both granular (RS2) and retrograded (RS3) form, are among the most used commercial RS ingredients in bread-making. They are widely available from many companies including Ingredion, Roquette, Cerestar, and SunOpta Ingredients. Tapioca rich in retrograded amylose (RS3) has also been investigated, which can be purchased from Cargill. RS2 from green banana starch or flour has also been evaluated in bread. The demand for banana starch/flour (RS2) is on the rise, and companies like Chiquita (Costa Rica), Livekuna (Canada), International Agriculture Group (United States), and Natural Evolution (Australia) commercialize a wide range of banana starch/flour ingredients with elevated RS content (~40–50%). On the one hand, it must be brought into attention that, converse to RS2 from high-amylose maize and RS3, RS2 from green bananas is not heat-stable and will not resist baking, that is, there will be a significant fraction of RS that will be lost during baking [9,122]. It is noteworthy that banana RS2 decreases with ripening owing to its conversion into reducing sugars by endogenous α-amylase [123]. This enzyme has been reported to present an optimal activity between 8 °C and 38 °C, starting to be denatured at 38 °C and being fully denatured after 5 min at 100 °C [124]. Therefore, the drying step will be critical for its inactivation and the preservation of RS2 in banana flours. Specifically, Pico et al. [125] showed how oven-dried banana flours at 40 °C for 24 h exhibited an insoluble dietary fiber content of 26.8%, which was significantly lower than the same flours obtained through freeze-drying (43.3%). On the other hand, albeit banana RS is lost during baking, banana starch has been reported to have a suitable molecular structure to result in structurally-driven slowly digestible starch in bread crumb after baking through retrogradation [9], part of which could reach the colon as RS3 (RS3b). This occurrence has been reported to improve through shear-induced fragmentation of amylopectin molecules through high shear extrusion [91], which was attributed by the authors to smaller amylopectin fragments being more mobile and more prone to interact through retrogradation.

The targeted amount of RS ingredient during formulation depends on the starch being used and the desired RS level in the bread. Normally, percentages of replacement have been reported within 5–30%, which resulted in breads with final RS content being dependent on the method of analysis. The RS content in both ingredients and breads was, in some cases, quantified by the AOAC Method 2002.02 and modifications of the Englyst procedure. However, AOAC official methods 985.29 and 991.43 were also used, which can lead to underestimations of the RS content (Section 2). Therefore, it is unfortunate that the RS content of different RS ingredients and breads is not comparable nor harmonized. It is expected that the CODEX definition of dietary fiber [126], and its adaptation by many worldwide authorities, brings about a unique method of analysis whose adoption enables harmonized information about the RS content in different commercial RS ingredients and RS-containing foods. This would also answer existing uncertainties in the association of RS consumption through breads with positive health outcomes. It is very important to mention that the clear majority of studies did not report information about the day RS was analyzed (i.e. time after baking), which masks information about the structural basis of the RS in breads, especially of those containing RS3, which increases over time through retrogradation [48,93,127–129]. Another masking factor is the fact that the nature of the sample for RS analysis is unknown and not reported in most studies, that is, whether crumb, crust, or the whole slice was analyzed (Table 2). This is particularly important considering the differences in the degree of gelatinization between crumb and crust (Figure 2) [6].

The incorporation of RS ingredients into bread formula also brings about differences in the physicochemical and organoleptic properties of the bread (Table 2). The flavor, mouth-feel, appearance, and texture are examples of important quality factors to bear in mind for good consumers' acceptance. As reported in Table 2, formulation of breads with increasing levels of RS2 and RS3 sources, in general, has detrimental effect on volume, hardness, cohesiveness, and crust color. On the basis of Table 2, approximately a 20% replacement of wheat flour by RS ingredients seems to be adequate to keep bread final quality, although lower specific volume and harder and/or less cohesive crumbs were generally observed. Paler crusts are visible in some studies owing to the whitish color of starch and the reduction in protein content available for Maillard reaction, while the color of the crumb seems less affected. Meanwhile, the incorporation of banana flour led to both darker crumb and crust [9]. Differences in crumb cell size distribution and decreased gluten network formation have also been reported [127,130–132] On the other hand, consumers' perception generally reflected similar or unaffected sensory evaluation, which may be because of the bland flavor of most RS sources. In this regard, Almeida et al. [133] studied the effects of adding different dietary fiber sources and concluded that RS2 (high amylose maize) was a more "inert" fiber source in relation to bread quality characteristics. RS2 was found to have lower water holding capacity than other dietary fibers, and thus less impact on dough rheology, resulting in breads with superior quality [134].

In addition, other than the nature of RS ingredients, processing conditions may also influence the formation of RS in bread. Baking under low-temperature and a long-time period significantly resulted in higher amounts of RS in bread than in those baked under higher-temperature and shorter time [48,49,135,136]. Similarly, higher addition of water in the formula has also been reported to increase RS in the bread. The higher the water content in the dough, the more starch can be gelatinized, resulting in increased starch retrogradation (RS3) during cooling of gelatinized starch [136]. RS3 in wheat bread has been reported to be greater for refrigeration than ambient or frozen temperatures [129], so for certain starches, refrigeration temperatures may boost their RS property in breads.

In some studies, the amount of water added to each formulation was adjusted based on the water binding capacity of starches as determined by farinographic analysis or elastic modulus [127,130,131,137–139]. In general, these RS rich ingredients have higher water absorption capacity than wheat flour or gluten-free flours used for bread making, especially if RS3, in non-granular form, is used [130,139]. Therefore, if water content is not properly adjusted, especially in gluten-containing breads, higher water absorption by RS ingredients in the dough can result in detriment of the gluten network [131,140]. Low water availability causes non-optimal repartition of water among dough components and may lead to final breads with detrimental quality characteristics in terms of specific volume, textural attributes, and appearance [141]. It is important to highlight that despite water adjustment in the formula, the specific volume always decreased when high levels of RS ingredients were added into the formulation of gluten-containing breads. This could be explained by the extent of gluten protein dilution [142] and a hindrance effect on the gluten network development by the non-gelatinized high maize starch granules [130]. Conversely, in gluten-free breads, no differences or even an improvement in bread volume with RS inclusion were observed in some studies [9,136,143].

Table 2. In vitro studies on commercially available RS2 and RS3 sources as ingredients to increase RS content in wheat- and gluten-free breads.

Ingredient	RS Content	Type of Bread	Substitution Level (%)	Evaluation Day	RS Content	In vitro RS Method	Effects on Bread Quality	Refs.
HA maize starch, Hi-Maize 260, Ingredion	60% IDF Manufacturer (58.4% TDF)	Wheat flour (4.5% TDF)	0 10 20 30	n.a. crumb	6.6% bb, db 9.5% bb, db 17.0% bb, db 26.6% bb, db	AOAC 985.29	Increased hardness Decreased cohesiveness and resilience Decreased volume with 30% level Lighter crust Decreased number of cells in the crumb Decreased C_{inf} and estimated GI Increased consumer acceptability (20% level)	[38]
HA maize starch, Hi-Maize 260, Ingredion	60% RS Manufacturer	Wheat flour	0 10 15 20	2 h, Lyophilized crumb	1.2% n.a. 3.9% n.a. 5.9% n.a. 11.1% n.a.	Goni et al. [40]	Decreased volume Increased hardness with 15% level Lighter crust Staling dependent on the level of replacement	[142]
HA, Amylo-maize starch N-400, Roquette	40% TDF Manufacturer	GF flour mix (maize starch, rice flour, tapioca starch)	0 20—RS2 20—RS3 [a]	n.a. bread	1.2% bb, db 4.4% bb, db 7.6–9.2% bb, db	Englyst et al. [31]	Reduced in vitro glycemic index (RS3 > RS2)	[137]
HA maize starch, Hi-Maize 260, Ingredion	56% TDF Manufacturer	Yellow maize flour	0 20	n.a.	4.3% bb, db 12.0% bb, db	Modified AOAC 2002.02	Specific volume and texture were not modified Decreased cell density Similar sensory evaluation SDS fraction was also increased eGI decreased from 85 to 71	[143]
HA maize starch, Hi-Maize 260, Ingredion	56% TDF Manufacturer	White maize flour	0 20	n.a.	5.5% bb, db 11.3% bb, db	Modified AOAC 2002.02	Specific volume and texture were not modified Same cell density Similar sensory evaluation SDS fraction was also increased eGI decreased from 83 to 72	[143]
HA maize starch, Eurylon, Roquette	83.2% RS2	Wheat flour (14.1% RS)	0 20 [b]	24 h/7 days, Lyophilized crumb	0.0%/4.4% bb, db 7.7%/10.2% bb, db	Modified Englyst et al. [31]	Decreased specific volume Decreased hardness No sensory differences	[127]
Extruded retrograded HA maize starch, EURESTA, Cerestar	29.5% RS3	Wheat flour (14.1% RS)	0 20 [b]	24 h/7 days, Lyophilized crumb	0.0%/4.4% bb, db 8.4%/11.0% bb, db	Modified Englyst et al. [31]	Decreased specific volume Decreased hardness No sensory differences	[127]
HA maize starch, HylonVII, Ingredion	53% RS2	Wheat flour	0 10 20 30	24 h, Lyophilized crumb	1.2% bb, db 4.1% bb, db 8.1% bb, db 10.1% bb, db	AOAC 2002.02	Decreased volume for 30% level Increased hardness for 30% level Paler crust color for 20% and 30% levels	[139]

Table 2. Cont.

Ingredient	RS Content	Type of Bread	Substitution Level (%)	Evaluation Day	RS Content	In vitro RS Method	Effects on Bread Quality	Refs.
HA maize starch, Novelose330, Ingredion	46.5% RS3	Wheat flour	0 10 20 30	24 h, Lyophilized crumb	1.2% bb, db 4.7% bb, db 9.7% bb, db 12.7% bb, db	AOAC 2002.02	Decreased volume 20% and 30% levels Increased hardness for 20% and 30% levels Paler crust color for 30%	[139]
HA maize starch CrystaLean, SunOpta ingredients	45% RS3	Wheat flour	0 10 20 30	24 h, Lyophilized crumb	1.2% bb, db 4.4% bb, db 8.3% bb, db 12.4% bb, db	AOAC 2002.02	Decreased volume for 30% level Increased hardness for 30% Paler crust color above 30%	[139]
HA maize starch, Hi-Maize 260, Ingredion	RS > 60% Manufacturer	Maize starch Potato starch (4:1) [c]	0 10 15 20	n.a.	2.1% bb 3.9% bb 4.7% bb 5.0% bb	AOAC 991.43	Decreased volume Similar initial hardness (20% level) Reduced hardening (48 h)	[138]
Tapioca starch, ActiStar 11700, Cargill	RS3 > 50% Manufacturer	Maize starch: Potato starch (4:1 mixture) [d]	0 10 15 20	n.a.	2.1% bb 2.5% bb 2.8% bb 3.0% bb	AOAC 991.43	Decreased volume Reduced initial hardness (20% level) Reduced hardening (48 h)	[138]
HA wheat flour, Okumoto Flour milling	6.7% TDF	Wheat flour (3.4% TDF)	0 10 30 50	2h	0.9% bb, db 1.6% bb, db 2.6% bb, db 3.0% bb, db	AOAC 985.29 [e]	Decreased volume Increased hardness Similar staling (hardness) for 30 and 50% levels Higher staling for 10% level Increasing RS with storage and substitution	[132]
Green banana starch, Natural Evolution	42.2% RS2	Maize starch (0.8% RS): Rice flour (0.1% RS) (1:1 mixture)	0 20—Native 20—Extruded [f]	24 h, Crumb and crust	1.5% cb − 0.3% ct, db 1.7% cb −5.7% ct, db 1.9% cb −0.7% ct, db	AOAC 2002.02	Darker bread color Improved volume, reduced hardness, and improved sensory acceptance (native banana) Increased SDS fraction in crumb with native and extruded banana	[9]
Green plantain flour, Chiquita	50.1% RS2	Rice flour: GF wheat starch (1:1 mixture)	0 35	24 h	1.1% bb, db 2.3% bb, db	AOAC 2002.02	Improved volume but increased firmness Darker bread crumb Lighter bread crust Optimization of water content, baking time and temperature for 30% replacement to maximize RS (3%)	[136]

SDS = slowly digestible starch; AOAC = Association of Official Analytical Chemists; IDF = insoluble dietary fiber; TDF = total dietary fiber; GI = glycemic index; GF = gluten-free; n.a. = not available; HA = high amylose; bb = bread basis; cb = crumb basis; ct = crust basis; db = dry basis. [a] RS3 was prepared with RS from HA maize starch (Amylomaize, N400) subjected to debranching and/or three autoclaving–cooling cycles. [b] Wheat flour (24%) was replaced by 20% of RS2 or RS3 source and 4% of gluten. [c] Maize starch was replaced by HA maize starch. Water level was not modified in the formula. [d] Potato starch was replaced by tapioca starch. Water level was increased in the formula. [e] TDF and RS in breads after baking and during storage were determined using the total dietary fiber assay kit. The RS in bread was calculated as the amount of non-digestible carbohydrate minus DF that already existed in the flours. [f] Banana starch (Native RS2) was extruded under high-shear extrusion to obtain pregelatinized starch (RS3).

5. Current Legislation of RS Ingredients and Products in the Food Industry

RS generally meets the criteria to be defined as "dietary fiber" by the comprehensive dietary fiber definition adopted by the CODEX Alimentarius Commission [126]. However, isolated or synthetic RS ingredients require the American Food and Drug Administration (FDA) [144] or European Union (EU) approval [145] after assessments of scientific evidence relating RS to physiological benefits. Under that proposed outline, "isolated" (pure RS2) or "synthetic" (RS3, RS4 and RS5) RS sources would remain outside this definition. Nonetheless, in June 2018, the FDA [144] released a review of the scientific evidence on the physiological effects of certain non-digestible carbohydrates, which decided to include isolated RS2 ingredients, such as raw green banana, potato, and high amylose starches, in the definition of dietary fiber. According to this categorization, in Europe (2 kcal/g), Australia (2 kcal/g), and USA (0 kcal/g), RS has a lower energy value compared with other non-fiber carbohydrates (4 kcal/g) [23,146]. The European Commission [147] also allows manufactures to voluntarily claim foods as a "source of fiber" if it contains at least 3 g of fiber per 100 g, and as "high in fiber" if it contains at least 6 g of fiber per 100 g.

Current regulations also identify the potential physiological benefits of RS. The European Food and Safety Authority (EFSA) approved the health claim, "Replacing digestible starch with resistant starch induces a lower blood glucose rise after a meal". However, this claim can be only used when the final RS content in the food is at least 14% of the total starch [148]. On the other hand, from 2016, the FDA [144] has allowed manufacturers to use the claim related to high amylose maize RS, "High-amylose maize resistant starch, a type of fiber, may reduce the risk of type 2 diabetes, although FDA has concluded that there is limited scientific evidence for this claim". So far, to the best of our knowledge, there is no other RS source with an authorized health claim in the United States.

6. Conclusions

The development of breads rich in RS and acceptable quality attributes could have a positive impact on the modulation of the glycemic response, the control of body weight, and the improvement of bowel health of bread consumers. The growing evidence of the positive health outcomes attributed to RS is leading to the apparition of novel RS ingredients in the market for bread-making, whose incorporation may seem the most logical and easy strategy to increase the RS content in breads. However, it must be noted that not all RS ingredients preserve the RS property during baking. It is thus paramount to understand the structural basis for their resistance to digestion and hydrothermal processing, which is often disregarded. This review concludes that high amylose starches, both native (RS2) and processed (RS3), are the most suitable RS ingredients for bread making in terms of RS preservation during baking and a lower detrimental impact on bread texture. However, their level of inclusion must be carefully selected. Another issue that this review addresses is the lack of harmony in RS values, which is the result of using different in vitro methods, some of which do not account for all types of RS structures. This outlook is changing though, as AOAC Method 2002.02 or any of its extensions, such as AOAC Methods 2009.01/2011.25, are adopted by many researchers from different nationalities.

Author Contributions: Conceptualization, L.R. and M.M.M.; Writing—original draft, L.R. and M.M.M.; Writing—review, M.M.M.

Funding: This research was funded by the Natural Sciences and Engineering Research Council of Canada (NSERC) Discovery program, grant number 401499.

Conflicts of Interest: The authors declare no conflict of interest. The funder had no role in the design of the study, the interpretation of data, and the writing of the manuscript.

References

1. Lafiandra, D.; Riccardi, G.; Shewry, P.R. Improving cereal grain carbohydrates for diet and health. *J. Cereal Sci.* **2014**, *59*, 312–326. [CrossRef] [PubMed]
2. Cust, A.E.; Skilton, M.R.; van Bakel, M.M.; Halkjaer, J.; Olsen, A.; Agnoli, C.; Psaltopoulou, T.; Buurma, E.; Sonestedt, E.; Chirlaque, M.D.; et al. Total dietary carbohydrate, sugar, starch and fibre intakes in the European Prospective Investigation into Cancer and Nutrition. *Eur. J. Clin. Nutr.* **2009**, *63*, S37–S60. [CrossRef] [PubMed]
3. Aune, D.; Keum, N.; Giovannucci, E.; Fadnes, L.T.; Boffetta, P.; Greenwood, D.C.; Norat, T. Whole grain consumption and risk of cardiovascular disease, cancer, and all cause and cause specific mortality: Systematic review and dose-response meta-analysis of prospective studies. *BMJ* **2016**, *353*, i2716. [CrossRef] [PubMed]
4. Pot, G.K.; Prynne, C.J.; Almoosawi, S.; Kuh, D.; Stephen, A.M. Trends in food consumption over 30 years: Evidence from a British birth cohort. *Eur. J. Clin. Nutr.* **2015**, *69*, 817. [CrossRef] [PubMed]
5. Sajilata, M.G.; Singhal, R.S.; Kulkarni, P.R. Resistant starch–a review. *Compr. Rev. Food Sci. F.* **2006**, *5*, 1–17. [CrossRef]
6. Martinez, M.M.; Roman, L.; Gomez, M. Implications of hydration depletion in the in vitro starch digestibility of white bread crumb and crust. *Food Chem.* **2018**, *239*, 295–303. [CrossRef] [PubMed]
7. Thorvaldsson, K.; Skjöldebrand, C. Water diffusion in bread during baking. *LWT* **1998**, *31*, 658–663. [CrossRef]
8. Primo-Martín, C.; van Nieuwenhuijzen, N.H.; Hamer, R.J.; van Vliet, T. Crystallinity changes in wheat starch during the bread-making process: Starch crystallinity in the bread crust. *J. Cereal Sci.* **2007**, *45*, 219–226. [CrossRef]
9. Roman, L.; Gomez, M.; Hamaker, B.R.; Martinez, M.M. Banana starch and molecular shear fragmentation dramatically increase structurally driven slowly digestible starch in fully gelatinized bread crumb. *Food Chem.* **2019**, *274*, 664–671. [CrossRef]
10. Ceriello, A.; Colagiuri., S. International Diabetes Federation guideline for management of postmeal glucose: A review of recommendations. *Diabetic Med.* **2008**, *25*, 1151–1156. [CrossRef]
11. Brand-Miller, J.; McMillan-Price, J.; Steinbeck, K.; Caterson, I. Dietary glycemic index: Health implications. *J. Am. Coll. Nutr.* **2009**, *28*, 446S–449S. [CrossRef] [PubMed]
12. Jenkins, D.J.; Kendall, C.W.; Augustin, L.S.; Franceschi, S.; Hamidi, M.; Marchie, A.; Jenkins, A.L.; Axelsen, M. Glycemic index: Overview of implications in health and disease. *Am. J. Clin. Nutr.* **2002**, *76*, 266S–273S. [CrossRef] [PubMed]
13. Reynolds, A.; Mann, J.; Cummings, J.; Winter, N.; Mete, E.; Te Morenga, L. Carbohydrate quality and human health: A series of systematic reviews and meta-analyses. *The Lancet* **2019**, *393*, 434–445. [CrossRef]
14. Bindels, L.B.; Walter, J.; Ramer-Tait, A.E. Resistant starches for the management of metabolic diseases. *Curr. Opin. Clin. Nutr.* **2015**, *18*, 559–565. [CrossRef] [PubMed]
15. Birt, D.F.; Boylston, T.; Hendrich, S.; Jane, J.-L.; Hollis, J.; Li, L.; McClelland, J.; Moore, S.; Phillips, G.J.; Schalinske, M.R.K.; et al. Resistant Starch: Promise for Improving Human Health. *Adv. Nutr.* **2013**, *4*, 587–601. [CrossRef]
16. Bird, A.R.; Lopez-Rubio, A.; Shrestha, A.K.; Gidley, M.J. Resistant starch in vitro and in vivo: Factors determining yield, structure, and physiological relevance. In *Modern Biopolymer Science*; Kasapis, S., Ubbink, J., Eds.; IT Norton, Academic Press: London, UK, 2009; pp. 449–510.
17. Bird, A.; Conlon, M.; Christophersen, C.; Topping, D. Resistant starch, large bowel fermentation and a broader perspective of prebiotics and probiotics. *Benef. Microbes* **2010**, *1*, 423–431. [CrossRef]
18. Dupuis, J.H.; Liu, Q.; Yada, R.Y. Methodologies for increasing the resistant starch content of food starches: A review. *Compr. Rev. Food Sci. F.* **2014**, *13*, 1219–1234. [CrossRef]
19. Fuentes-Zaragoza, E.; Sánchez-Zapata, E.; Sendra, E.; Sayas, E.; Navarro, C.; Fernández-López, J.; Pérez-Alvarez, J.A. Resistant starch as prebiotic: A review. *Starch-Stärke* **2011**, *63*, 406–415. [CrossRef]
20. Higgins, J.A.; Brown, I.L. Resistant starch: A promising dietary agent for the prevention/treatment of inflammatory bowel disease and bowel cancer. *Curr. Opin. Gastroenterol.* **2013**, *29*, 190–194. [CrossRef]
21. Keenan, M.J.; Zhou, J.; Hegsted, M.; Pelkman, C.; Durham, H.A.; Coulon, D.B.; Martin, R.J. Role of resistant starch in improving gut health, adiposity, and insulin resistance. *Adv. Nutr.* **2015**, *6*, 198–205. [CrossRef]
22. Koh, G.Y.; Rowling, M.J. Resistant starch as a novel dietary strategy to maintain kidney health in diabetes mellitus. *Nutr. Rev.* **2017**, *75*, 350–360. [CrossRef] [PubMed]

23. Lockyer, S.; Nugent, A.P. Health effects of resistant starch. *Nutr. Bull.* **2017**, *42*, 10–41. [CrossRef]
24. Qi, X.; Tester, R.F. Utilisation of dietary fibre (non-starch polysaccharide and resistant starch) molecules for diarrhoea therapy: A mini-review. *Int. J. Biol. Macromol.* **2019**, *122*, 572–577. [CrossRef] [PubMed]
25. Raigond, P.; Ezekiel, R.; Raigond, B. Resistant starch in food: A review. *J. Sci. Food Agric.* **2015**, 1968–1978. [CrossRef] [PubMed]
26. Robertson, M.D.; Bickerton, A.S.; Dennis, A.L.; Vidal, H.; Frayn, K.N. Insulin-sensitizing effects of dietary resistant starch and effects on skeletal muscle and adipose tissue metabolism. *Am. J. Clin. Nutr.* **2005**, *82*, 559–567. [CrossRef] [PubMed]
27. Yuan, H.C.; Meng, Y.; Bai, H.; Shen, D.Q.; Wan, B.C.; Chen, L.Y. Meta-analysis indicates that resistant starch lowers serum total cholesterol and low-density cholesterol. *Nutr. Res.* **2018**, *54*, 1–11. [CrossRef] [PubMed]
28. Champ, M. Resistant starch. In *Starch in Food: Structure, Function and Applications*; Eliasson, A.C., Ed.; CRC Press: New York, NY, USA, 2004; pp. 560–574.
29. Berry, C.S. Resistant starch formation: Formation and measurement of starch that survives exhaustive digestion with amylolytic enzymes during the determination of dietary fibre. *J. Cereal Sci.* **1986**, *4*, 301–314. [CrossRef]
30. Englyst, H.; Wiggins, H.S.; Cummings, J.H. Determination of the non-starch polysaccharides in plant foods by gas-liquid chromatography of constituent sugars as alditol acetates. *Analyst* **1982**, *107*, 307–318. [CrossRef]
31. Englyst, H.N.; Kingman, S.M.; Cummings, J.H. Classification and measurement of nutritionally important starch fractions. *Eur. J. Clin. Nutr.* **1992**, *46*, S33–S50.
32. Englyst, H.N.; Kingman, S.M.; Hudson, G.J.; Cummings, J.H. Measurement of resistant starch in vitro and in vivo. *Br. J. Nutr.* **1996**, *75*, 749–755. [CrossRef]
33. Englyst, K.N.; Hudson, G.J.; Englyst, H.N. Starch analysis in food. In *Encyclopedia of Analytical Chemistry*; Meyers, R.A., Ed.; John Wiley & Sons: Chichester, UK, 2000; pp. 4246–4262.
34. McCleary, B.V.; Monaghan, D.A. Measurement of resistant starch. *J. AOAC Int.* **2002**, *85*, 665–675. [PubMed]
35. AOAC. *AOAC Official Methods of Analysis*, 17th ed.; Association of Official Analytical Chemists International: Gaithersburg, MD, USA, 2000.
36. AACC International. *AACC International Approved Methods*; AACC International: St. Paul, MN, USA, 2015.
37. Akerberg, A.K.; Liljeberg, H.G.; Granfeldt, Y.E.; Drews, A.W.; Björck, I.M. An in vitro method, based on chewing, to predict resistant starch content in foods allows parallel determination of potentially available starch and dietary fiber. *J. Nutr.* **1998**, *128*, 651–660. [CrossRef] [PubMed]
38. Champ, M. Determination of resistant starch in foods and food products: Interlaboratory study. *Eur. J. Clin. Nutr.* **1992**, *46*, S51–S62. [PubMed]
39. Champ, M.; Martin, L.; Noah, L.; Gratas, M. Analytical methods for resistant starch. In *Complex Carbohydrates in Foods*; Cho, S.S., Prosky, L., Dreher, M., Eds.; Marcel Dekker Inc: New York, NY, USA, 1999; pp. 169–187.
40. Goni, I.; Garcia-Diz, L.; Manas, E.; Saura-Calixto, F. Analysis of Resistant Starch: A Method for Foods and Food Products. *Food Chem.* **1996**, *56*, 445–449. [CrossRef]
41. Muir, J.G.; O'Dea, K. Validation of an in vitro assay for predicting the amount of starch that escapes digestion in the small intestine of humans. *Am. J. Clin. Nutr.* **1993**, *57*, 540–546. [CrossRef] [PubMed]
42. Prosky, L.; Asp, N.G.; Furda, I.; DeVries, J.W.; Schweizer, T.F.; Harland, B.F. Determination of total dietary fiber in foods and food products: Collaborative study. *J. Assoc. Off. Anal. Chem.* **1985**, *68*, 677–679. [PubMed]
43. Lee, S.C.; Prosky, L. International survey on dietary fiber: Definition, analysis, and reference materials. *J. AOAC Int.* **1995**, *78*, 22–36.
44. McCleary, B.V. An integrated procedure for the measurement of total dietary fibre (including resistant starch), non-digestible oligosaccharides and available carbohydrates. *Anal. Bioanal. Chem.* **2007**, *389*, 291–308. [CrossRef]
45. McCleary, B.V.; De Vries, J.W.; Rader, J.I.; Cohen, G.; Prosky, L.; Mugford, D.C.; Champ, M.; Okuma, K. Determination of total dietary fiber (CODEX definition) by enzymatic-gravimetric method and liquid chromatography: Collaborative study. *J. AOAC Int.* **2010**, *93*, 221–233. [CrossRef]
46. Eerlingen, R.C.; Delcour, J.A. Formation, analysis, structure, and properties of type III enzyme resistant starch. *J. Cereal Sci.* **1995**, *22*, 129–138. [CrossRef]
47. Brown, I.L.; Yotsuzuka, M.; Birkett, A.; Henriksson, A. Prebiotics, synbiotics and resistant starch. *J. Japanese. Assoc. Dietary Fiber Res.* **2006**, *10*, 1–9.

48. Amaral, O.; Guerreiro, C.S.; Gomes, A.; Cravo, M. Resistant starch production in wheat bread: Effect of ingredients, baking conditions and storage. *Eur Food Res Technol.* **2016**, *242*, 1747–1753. [CrossRef]
49. Liljeberg, H.; Akerberg, A.; Bjorck, I. Resistant starch formation in bread as influenced by choice of ingredients or baking conditions. *Food Chem.* **1996**, *56*, 389–394. [CrossRef]
50. Wen, Q.B.; Lorenz, K.J.; Martin, D.J.; Stewart, B.G.; Sampson, D.A. Carbohydrate Digestibility and Resistant Starch of Steamed Bread. *Starch-Stärke* **1996**, *48*, 180–185. [CrossRef]
51. Dhital, S.; Gidley, M.J.; Warren, F.J. Inhibition of α-amylase activity by cellulose: Kinetic analysis and nutritional implications. *Carbohydr. Polym.* **2015**, *123*, 305–312. [CrossRef] [PubMed]
52. Bhattarai, R.R.; Dhital, S.; Mense, A.; Gidley, M.J.; Shi, Y.C. Intact cellular structure in cereal endosperm limits starch digestion in vitro. *Food. Hydrocoll.* **2018**, *81*, 139–148. [CrossRef]
53. Bhattarai, R.R.; Dhital, S.; Wu, P.; Chen, X.D.; Gidley, M.J. Digestion of isolated legume cells in a stomach-duodenum model: Three mechanisms limit starch and protein hydrolysis. *Food Funct.* **2017**, *8*, 2573–2582. [CrossRef] [PubMed]
54. Dhital, S.; Bhattarai, R.R.; Gorham, J.; Gidley, M.J. Intactness of cell wall structure controls the in vitro digestion of starch in legumes. *Food Funct.* **2016**, *7*, 1367–1379. [CrossRef]
55. Edwards, C.H.; Grundy, M.M.; Grassby, T.; Vasilopoulou, D.; Frost, G.S.; Butterworth, P.J.; Berry, S.E.E.; Sanderson, J.; Ellis, P.R. Manipulation of starch bioaccessibility in wheat endosperm to regulate starch digestion, postprandial glycemia, insulinemia, and gut hormone responses: A randomized controlled trial in healthy ileostomy participants. *Am. J. Clin. Nutr.* **2015**, *102*, 791–800. [CrossRef]
56. Livesey, G.; Wilkinson, J.A.; Roe, M.; Faulks, R.; Clark, S.; Brown, J.C.; Kennedy, H.; Elia, M. Influence of the physical form of barley grain on the digestion of its starch in the human small intestine and implications for health. *Am. J. Clin. Nutr.* **1995**, *61*, 75–81. [CrossRef]
57. Roman, L.; Gomez, M.; Li, C.; Hamaker, B.R.; Martinez, M.M. Biophysical features of cereal endosperm that decrease starch digestibility. *Carbohydr. Polym.* **2017**, *165*, 180–188. [CrossRef] [PubMed]
58. Birkett, A.M.; Brown, I.L. Resistant starch and health. In *Technology of Functional Cereal Products*; Hamaker, B.R., Ed.; CRC Press: West Palm Beach, FL, USA, 2008; pp. 63–85.
59. Al-Rabadi, G.J.S.; Gilbert, R.G.; Gidley, M.J. Effect of particle size on kinetics of starch digestion in milled barley and sorghum grains by porcine alpha amylase. *J. Cereal Sci.* **2009**, *50*, 198–204. [CrossRef]
60. Granfeldt, Y.; Liljeberg, H.; Drews, A.; Newman, R.; Björck, I. Glucose and insulin responses to barley products: Influence of food structure and amylose-amylopectin ratio. *Am. J.Clin..Nutr.* **1994**, *59*, 1075–1082. [CrossRef] [PubMed]
61. Jenkins, D.J.; Wesson, V.; Wolever, T.M.; Jenkins, A.L.; Kalmusky, J.; Guidici, S.; Csima, A.; Josse, R.G.; Wong, G.S. Wholemeal versus wholegrain breads: Proportion of whole or cracked grain and the glycaemic response. *BMJ* **1988**, *297*, 958–960. [CrossRef] [PubMed]
62. Martinez, M.M.; Calviño, A.; Rosell, C.M.; Gomez, M. Effect of different extrusion treatments and particle size distribution on the physicochemical properties of rice flour. *Food Bioprocess.Tech.* **2014**, *7*, 2657–2665. [CrossRef]
63. De La Hera, E.; Rosell, C.M.; Gomez, M. Effect of water content and flour particle size on gluten-free bread quality and digestibility. *Food Chem.* **2014**, *151*, 526–531. [CrossRef]
64. Protonotariou, S.; Mandala, I.; Rosell, C.M. Jet milling effect on functionality, quality and in vitro digestibility of whole wheat flour and bread. *Food Bioprocess. Tech.* **2015**, *8*, 1319–1329. [CrossRef]
65. Ezeogu, L.I.; Duodu, K.G.; Taylor, J.R.N. Effects of endosperm texture and cooking conditions on the in vitro starch digestibility of sorghum and maize flours. *J. Cereal. Sci.* **2005**, *42*, 33–44. [CrossRef]
66. Zhang, G.; Hamaker, B.R. Low α-amylase starch digestibility of cooked sorghum flours and the effect of protein. *Cereal. Chem.* **1998**, *75*, 710–713. [CrossRef]
67. Berti, C.; Riso, P.; Monti, L.D.; Porrini, M. In vitro starch digestibility and in vivo glucose response of gluten-free foods and their gluten counterparts. *Eur. J. Nutr.* **2004**, *43*, 198–204. [CrossRef]
68. Jenkins, D.J.; Thorne, M.J.; Wolever, T.M.; Jenkins, A.L.; Rao, A.V.; Thompson, L.U. The effect of starch-protein interaction in wheat on the glycemic response and rate of in vitro digestion. *Am. J. Clin. Nutr.* **1987**, *45*, 946–951. [CrossRef] [PubMed]
69. Taylor, J.R.; Taylor, J.; Campanella, O.H.; Hamaker, B.R. Functionality of the storage proteins in gluten-free cereals and pseudocereals in dough systems. *J. Cereal. Sci.* **2016**, *67*, 22–34. [CrossRef]

70. Adams, C.A.; Novellie, L.; Liebenberg, N.V.D.W. Biochemical properties and ultrastructure of protein bodies isolated from selected cereals. *Cereal. Chem.* **1976**, *53*, 1–12.
71. Batterman-Azcona, S.J.; Lawton, J.W.; Hamaker, B.R. Effect of specific mechanical energy on protein bodies and a-zeins in corn flour extrudates. *Cereal. Chem.* **1999**, *76*, 316–320. [CrossRef]
72. Batterman-Azcona, S.J.; Hamaker, B.R. Changes occurring in protein body structure and a-zein during cornflake processing. *Cereal. Chem.* **1998**, *75*, 217–221. [CrossRef]
73. Vamadevan, V.; Bertoft, E. Structure-function relationships of starch components. *Starch-Stärke* **2015**, *67*, 55–68. [CrossRef]
74. Inan Eroglu, E.; Buyuktuncer, Z. The effect of various cooking methods on resistant starch content of foods. *Nutr. Food Sci.* **2017**, *47*, 522–533. [CrossRef]
75. Faisant, N.; Buleon, A.; Colonna, P.; Molis, C.; Lartigue, S.; Galmiche, J.P.; Champ, M. Digestion of raw banana starch in the small intestine of healthy humans: Structural features of resistant starch. *Br. J. Nutr.* **1995**, *73*, 111–123. [CrossRef]
76. Takeda, C.; Takeda, Y.; Hizukuri, S. Structure of the amylopectin fraction of amylomaize. *Carbohydr. Res.* **1993**, *246*, 273–281. [CrossRef]
77. Planchot, V.; Colonna, P. Purification and characterization of extracellular alpha-amylase from Aspergillus fumigatus. *Carbohydr. Res.* **1995**, *272*, 97–109. [CrossRef]
78. Gallant, D.J.; Bouchet, B.; Baldwin, P.M. Microscopy of starch: Evidence of a new level of granule organization. *Carbohydr. Polym.* **1997**, *32*, 177–191. [CrossRef]
79. Dhital, S.; Warren, F.J.; Butterworth, P.J.; Ellis, P.R.; Gidley, M.J. Mechanisms of starch digestion by α-amylase—Structural basis for kinetic properties. *Crit. Rev. Food Sci. Nutr.* **2017**, *57*, 875–892. [CrossRef] [PubMed]
80. Haralampu, S. Resistant starch–a review of the physical properties and biological impact of RS3. *Carbohydr. Polym.* **2000**, *41*, 285–292. [CrossRef]
81. Jacobs, H.; Delcour, J. Hydrothermal modifications of granular starch with retention of the granular structure: A review. *J. Agric. Food Chem.* **1998**, *46*, 2895–2905. [CrossRef]
82. Htoon, A.; Shrestha, A.K.; Flanagan, B.M.; Lopez-Rubio, A.; Bird, A.R.; Gilbert, E.P.; Gidley, M.J. Effects of processing high amylose maize starches under controlled conditions on structural organization and amylase digestibility. *Carbohydr. Polym.* **2009**, *75*, 236–245. [CrossRef]
83. Patel, H.; Royall, P.G.; Gaisford, S.; Williams, G.R.; Edwards, C.H.; Warren, F.J.; Flanagan, B.M.; Ellis, P.R.; Butterworth, P.J. Structural and enzyme kinetic studies of retrograded starch: Inhibition of α-amylase and consequences for intestinal digestion of starch. *Carbohydr. Polym.* **2017**, *164*, 154–161. [CrossRef] [PubMed]
84. Englyst, H.N.; Cummings, J.H. Digestion of the polysaccharides of some cereal foods in the human small intestine. *Am. J. Clin. Nutr.* **1985**, *42*, 778–787. [CrossRef]
85. Vasanthan, T.; Bhatty, R.S. Enhancement of resistant starch (RS3) in amylomaize, barley, field pea and lentil starches. *Starch-Stärke* **1998**, *50*, 286–289. [CrossRef]
86. Lopez-Rubio, A.; Flanagan, B.M.; Shrestha, A.K.; Gidley, M.J.; Gilbert, E.P. Molecular rearrangement of starch during in vitro digestion: Toward a better understanding of enzyme resistant starch formation in processed starches. *Biomacromolecules* **2008**, *9*, 1951–1958. [CrossRef]
87. Cairns, P.; Sun, L.; Morris, V.J.; Ring, S.G. Physicochemical studies using amylose as an in vitro model for resistant starch. *J. Cereal Sci.* **1995**, *21*, 37–47. [CrossRef]
88. Gidley, M.J.; Cooke, D.; Darke, A.H.; Hoffmann, R.A.; Russell, A.L.; Greenwell, P. Molecular order and structure in enzyme-resistant retrograded starch. *Carbohydr. Polym.* **1995**, *28*, 23–31. [CrossRef]
89. Lopez-Rubio, A.; Htoon, A.; Gilbert, E.P. Influence of extrusion and digestion on the nanostructure of high-amylose maize starch. *Biomacromolecules* **2007**, *8*, 1564–1572. [CrossRef] [PubMed]
90. Zhang, B.; Dhital, S.; Flanagan, B.M.; Luckman, P.; Halley, P.J.; Gidley, M.J. Extrusion induced low-order starch matrices: Enzymic hydrolysis and structure. *Carbohydr. Polym.* **2015**, *134*, 485–496. [CrossRef] [PubMed]
91. Roman, L.; Campanella, O.; Martinez, M.M. Shear-induced molecular fragmentation decreases the bioaccessibility of fully gelatinized starch and its gelling capacity. *Carbohydr. Polym.* **2019**, *215*, 198–206. [CrossRef] [PubMed]
92. Zhang, G.; Sofyan, M.; Hamaker, B.R. Slowly digestible state of starch: Mechanism of slow digestion property of gelatinized maize starch. *J. Agric. Food Chem.* **2008**, *56*, 4695–4702. [CrossRef] [PubMed]

93. Martinez, M.M.; Li, C.; Okoniewska, M.; Mukherjee, N.; Vellucci, D.; Hamaker, B.R. Slowly digestible starch in fully gelatinized material is structurally driven by molecular size and A and B1 chain lengths. *Carbohydr. Polym.* **2018**, *197*, 531–539. [CrossRef] [PubMed]
94. Englyst, H.N.; Macfarlane, G.T. Breakdown of resistant and readily digestible starch by human gut bacteria. *J. Sci. Food Agric.* **1986**, *37*, 699–706. [CrossRef]
95. Benmoussa, M.; Moldenhauer, K.A.; Hamaker, B.R. Rice amylopectin fine structure variability affects starch digestion properties. *J. Agric. Food Chem.* **2007**, *55*, 1475–1479. [CrossRef]
96. Klucinec, J.D.; Thompson, D.B. Amylose and amylopectin interact in retrogradation of dispersed high-amylose starches. *Cereal. Chem.* **1999**, *76*, 282–291. [CrossRef]
97. Matalanis, AM.; Campanella, O.H.; Hamaker, B.R. Storage retrogradation behavior of sorghum, maize and rice starch pastes related to amylopectin fine structure. *J. Cereal Sci.* **2009**, *50*, 74–81. [CrossRef]
98. Maningat, C.C.; Seib, P.A. RS4-type resistant starch: Chemistry, functionality and health benefits. In *Resistant Starch Sources, Applications and Health Benefits*; Shi, Y.C., Maningat, C.C., Eds.; John Wiley: Hoboken, NJ, USA, 2013; pp. 43–77.
99. Ingredion. Official website. Technical documents for Versafibe 1490 from Ingredion website. Available online: https://www.ingredion.us/content/dam/ingredion/technical-documents/na/VERSAFIBE%201490%20%20%2006400400%20%20%20Technical%20Specification.pdf (accessed on 26 June 2019).
100. Roquette. Official website. Modified starches under the name "CLEARGUM". Available online: https://www.roquette.com/product-finder-food-nutrition/ (accessed on 25 June 2019).
101. MGP. Official website. Technical information on Fibersym RS4. Available online: https://www.mgpingredients.com/food-ingredients/products/line/fibersym (accessed on 26 June 2019).
102. Cargill. Official website. Technical information on RS4 starches C☆PolarTex, C☆StabiTex, C☆Tex. Available online: https://www.cargill.com/food-bev/emea/starches-derivatives/stabilized-starches (accessed on 26 June 2019).
103. Arp, C.G.; Correa, M.J.; Zuleta, Á.; Ferrero, C. Techno-functional properties of wheat flour-resistant starch mixtures applied to breadmaking. *Int. J. Food Sci. Technol.* **2017**, *52*, 550–558. [CrossRef]
104. Lee Yeo, L.; Seib, P.A. White pan bread and sugar-snap cookies containing wheat starch phosphate, a cross-linked resistant starch. *Cereal. Chem.* **2009**, *86*, 210–220.
105. Miller, R.A.; Bianchi, E. Effect of RS4 resistant starch on dietary fiber content of white pan bread. *Cereal. Chem.* **2017**, *94*, 185–189. [CrossRef]
106. Miyazaki, M.; Van Hung, P.; Maeda, T.; Morita, N. Recent advances in application of modified starches for breadmaking. *Trends. Food Sci. Technol.* **2006**, *17*, 591–599. [CrossRef]
107. Sciarini, L.S.; Bustos, M.C.; Vignola, M.B.; Paesani, C.; Salinas, C.N.; Perez, G.T. A study on fibre addition to gluten free bread: Its effects on bread quality and in vitro digestibility. *J. Food Sci. Technol.* **2017**, *54*, 244–252. [CrossRef] [PubMed]
108. Van Hung, P.; Morita, N. Dough properties and bread quality of flours supplemented with cross-linked corn starches. *Food Res. Int.* **2004**, *37*, 461–467. [CrossRef]
109. Wepner, B.; Berghofer, E.; Miesenberger, E.; Tiefenbacher, K.; NK Ng, P. Citrate starch—application as resistant starch in different food systems. *Starch-Stärke* **1999**, *51*, 354–361. [CrossRef]
110. Witczak, M.; Juszczak, L.; Ziobro, R.; Korus, J. Influence of modified starches on properties of gluten-free dough and bread. Part I: Rheological and thermal properties of gluten-free dough. *Food Hydrocoll.* **2012**, *28*, 353–360. [CrossRef]
111. Ziobro, R.; Korus, J.; Witczak, M.; Juszczak, L. Influence of modified starches on properties of gluten-free dough and bread. Part II: Quality and staling of gluten-free bread. *Food Hydrocoll.* **2012**, *29*, 68–74. [CrossRef]
112. Larsson, K.; Miezis, Y. On the possibility of dietary fiber formation by interaction in the intestine between starch and lipids. *Starch-Stärke* **1979**, *31*, 301–302. [CrossRef]
113. Morrison, W.R.; Law, R.V.; Snape, C.E. Evidence for Inclusion Complexes of Lipids with V-amylose in Maize, Rice and Oat Starches. *J. Cereal Sci.* **1993**, *18*, 107–109. [CrossRef]
114. Ai, Y.; Hasjim, J.; Jane, J. Effects of lipids on enzymatic hydrolysis and physical properties of starch. *Carbohydr. Polym.* **2013**, *92*, 120–127. [CrossRef] [PubMed]
115. Kawai, K.; Takato, S.; Sasaki, T.; Kajiwara, K. Complex formation, thermal properties, and in-vitro digestibility of gelatinized potato starch-fatty acid mixtures. *Food Hydrocoll.* **2012**, *27*, 228–234. [CrossRef]

116. Lau, E.; Zhou, W.; Henry, C.J. Effect of fat type in baked bread on amylose–lipid complex formation and glycaemic response. *Br. J. Nutr.* **2016**, *115*, 2122–2129. [CrossRef] [PubMed]
117. Biliaderis, C. Structural transitions and related physical properties of starch. In *Starch. Chemistry and Technology*; BeMiller, J., Whistler, R., Eds.; Academic Press: New York, NY, USA, 2009; pp. 293–372.
118. Schirmer, M.; Jekle, M.; Becker, T. Starch gelatinization and its complexity for analysis. *Starch-Stärke* **2015**, *67*, 30–41. [CrossRef]
119. Roman, L.; Belorio, M.; Gomez, M. Gluten-Free Breads: The Gap Between Research and Commercial Reality. *Compr. Rev. Food Sci. F.* **2019**, *18*, 690–702. [CrossRef]
120. Pareyt, B.; Finnie, S.M.; Putseys, J.A.; Delcour, J.A. Lipids in bread making: Sources, interactions, and impact on bread quality. *J. Cereal. Sci.* **2011**, *54*, 266–279. [CrossRef]
121. Hasjim, J.; Lee, SO.; Hendrich, S.; Setiawan, S.; Ai, Y.; Jane, J.L. Characterization of a novel resistant-starch and its effects on postprandial plasma-glucose and insulin responses. *Cereal. Chem.* **2010**, *87*, 257–262. [CrossRef]
122. Segundo, C.; Roman, L.; Gomez, M.; Martinez, M.M. Mechanically fractionated flour isolated from green bananas (M. cavendishii var. nanica) as a tool to increase the dietary fiber and phytochemical bioactivity of layer and sponge cakes. *Food Chem.* **2017**, *219*, 240–248. [CrossRef]
123. Segundo, C.; Roman, L.; Lobo, M.; Martinez, M.M.; Gomez, M. Ripe banana flour as a source of antioxidants in layer and sponge cakes. *Plant. Food Hum. Nutr.* **2017**, *72*, 365–371. [CrossRef]
124. Mao, W.W.; Kinsella, J.E. Amylase activity in banana fruit: Properties and changes in activity with ripening. *J. Food Sci.* **1981**, *46*, 1400–1403. [CrossRef]
125. Pico, J.; Xu, K.; Guo, M.; Mohamedshah, Z.; Ferruzzi, M.G.; Martinez, M.M. Manufacturing the ultimate green banana flour: Impact of drying and extrusion on phenolic profile and starch bioaccessibility. *Food Chem.* **2019**, *297*, 124990. [CrossRef] [PubMed]
126. CODEX, Joint FAO/WHO Food Standards Programme, Secretariat of the CODEX Alimentarius Commission: CODEX Alimentarius (CODEX) Guidelines on Nutrition Labeling CAC/GL 2–1985 as Last Amended 2010. Available online: http://www.fao.org/ag/humannutrition/33309-01d4d1dd1abc825f0582d9e5a2eda4a74.pdf (accessed on 26 June 2019).
127. Eerlingen, R.C.; Van Haesendonck, I.P.; De Paepe, G.; Delcour, J.A. Enzyme-resistant starch. III. The quality of straight-dough bread containing varying levels of enzyme-resistant starch. *Cereal. Chem.* **1994**, *71*, 165–169.
128. Niba, L.L. Effect of storage period and temperature on resistant starch and beta-glucan content in cornbread. *Food Chem.* **2003**, *83*, 493–498. [CrossRef]
129. Sullivan, W.R.; Hughes, J.G.; Cockman, R.W.; Small, D.M. The effects of temperature on the crystalline properties and resistant starch during storage of white bread. *Food Chem.* **2017**, *228*, 57–61. [CrossRef] [PubMed]
130. Arp, C.G.; Correa, M.J.; Ferrero, C. High-Amylose Resistant Starch as a Functional Ingredient in Breads: A Technological and Microstructural Approach. *Food Bioprocess. Tech.* **2018**, *11*, 2182–2193. [CrossRef]
131. Sanz-Penella, J.M.; Wronkowska, M.; Soral-Śmietana, M.; Collar, C.; Haros, M. Impact of the addition of resistant starch from modified pea starch on dough and bread performance. *Eur. Food Res. Technol.* **2010**, *231*, 499–508. [CrossRef]
132. Van Hung, P.; Yamamori, M.; Morita, N. Formation of Enzyme-Resistant Starch in Bread as Affected by High-Amylose Wheat Flour Substitutions. *Cereal. Chem. J.* **2005**, *82*, 690–694. [CrossRef]
133. Almeida, E.L.; Chang, Y.K.; Steel, C.J. Dietary fibre sources in bread: Influence on technological quality. *LWT* **2013**, *50*, 545–553. [CrossRef]
134. Sharma, A.; Yadav, B.S.; Ritika. Resistant starch: Physiological roles and food applications. *Food Rev. Int.* **2008**, *24*, 193–234. [CrossRef]
135. Akerberg, A.; Liljeberg, H.; Bjorck, I. Effects of amylose/amylopectin ratio and baking conditions on resistant starch formation and glycaemic indices. *J. Cereal. Sci.* **1998**, *28*, 71–80. [CrossRef]
136. Sarawong, C.; Gutierrez, Z.R.; Berghofer, E.; Schoenlechner, R. Effect of green plantain flour addition to gluten-free bread on functional bread properties and resistant starch content. *Int. J. Food Sci. Technol.* **2014**, *49*, 1825–1833. [CrossRef]
137. Giuberti, G.; Fortunati, P.; Gallo, A. Can different types of resistant starch influence the in vitro starch digestion of gluten free breads? *J. Cereal Sci.* **2016**, *100*, 253–255. [CrossRef]

138. Korus, J.; Witczak, M.; Ziobro, R.; Juszczak, L. The impact of resistant starch on characteristics of gluten-free dough and bread. *Food Hydrocoll.* **2009**, *23*, 988–995. [CrossRef]
139. Ozturk, S.; Koksel, H.; Ng, P.K. Farinograph properties and bread quality of flours supplemented with resistant starch. *Int. J. Food Sci. Nutr.* **2009**, *60*, 449–457. [CrossRef] [PubMed]
140. Goesaert, H.; Leman, P.; Delcour, J.A. Model approach to starch functionality in bread making. *J. Agric. Food Chem.* **2008**, *56*, 6423–6431. [CrossRef] [PubMed]
141. Tsatsaragkou, K.; Gounaropoulos, G.; Mandala, I. Development of gluten free bread containing carob flour and resistant starch. *LWT* **2014**, *58*, 124–129. [CrossRef]
142. Barros, J.H.; Telis, V.R.; Taboga, S.; Franco, C.M. Resistant starch: Effect on rheology, quality, and staling rate of white wheat bread. *J. Food Sci. Technol.* **2018**, *55*, 4578–4588. [CrossRef] [PubMed]
143. Collar, C.; Balestra, F.; Ancarani, D. Value added of resistant starch maize-based matrices in breadmaking: Nutritional and functional assessment. *Food Bioprocess. Tech.* **2014**, *7*, 3579–3590. [CrossRef]
144. FDA 2018. U.S. Department of Health and Human Services, Food and Drug Administration, Center for Food Safety and Applied Nutrition. 'Review of the scientific evidence on the physiological effects of non-digestible carbohydrates.'. June 2018. Available online: https://www.fda.gov/downloads/Food/LabelingNutrition/UCM610139.pdf. (accessed on 23 February 2019).
145. European Commission 2011. Regulation (EU) No 1169/2011 of the European Parliament and of the Council of 25 October 2011 on the provision of food information to consumers, amending Regulations (EC) No 1924/2006 and (EC) No 1925/2006 of the European Parliament and of the Council, and repealing Commission Directive 87/250/EEC, Council Directive 90/496/EEC, Commission Directive 1999/10/EC, Directive 2000/13/EC of the European Parliament and of the Council, Commission Directives 2002/67/EC and 2008/5/EC and Commission Regulation (EC) No 608/2004. Available online: http://eur-lex.europa.eu/legal-content/EN/TXT/PDF/?uri=CELEX:32011R1169&from=EN. (accessed on 23 February 2019).
146. Nugent, A.P. Health properties of resistant starch. *Nutr. Bull.* **2005**, *30*, 27–54. [CrossRef]
147. European Commission 2006. Regulation (EC) No 1924/2006 of the European Parliament and of the Council of 20 December 2006 on nutrition and health claims made on foods. Official Journal of the European Union, L404, 9–25. Available online: https://eur-lex.europa.eu/LexUriServ/LexUriServ.do?uri=OJ:L:2006:404:0009:0025:EN:PDF (accessed on 23 February 2019).
148. Stephen, A.M.; Champ, M.M.J.; Cloran, S.J.; Fleith, M.; van Lieshout, L.; Mejborn, H.; Burley, V.J. Dietary fibre in Europe: Current state of knowledge on definitions, sources, recommendations, intakes and relationships to health. *Nutr. Res. Rev.* **2017**, *30*, 149–190. [CrossRef] [PubMed]

© 2019 by the authors. Licensee MDPI, Basel, Switzerland. This article is an open access article distributed under the terms and conditions of the Creative Commons Attribution (CC BY) license (http://creativecommons.org/licenses/by/4.0/).

Review

Bovine Milk Fats and Their Replacers in Baked Goods: A Review

Zhiguang Huang [1,2], Letitia Stipkovits [1], Haotian Zheng [3], Luca Serventi [1] and Charles S. Brennan [1,2,*]

1. Department of Wine, Food and Molecular Biosciences, Faculty of Agriculture and Life Sciences, Lincoln University, Lincoln, Christchurch 7647, New Zealand
2. Riddet Research Institute, Palmerston North 4442, New Zealand
3. Dairy Innovation Institute, California Polytechnic State University, San Luis Obispo, CA 93407, USA
* Correspondence: Charles.Brennan@Lincoln.ac.nz; Tel.: +64-3-423-0637

Received: 8 July 2019; Accepted: 21 August 2019; Published: 2 September 2019

Abstract: Milk fats and related dairy products are multi-functional ingredients in bakeries. Bakeries are critical local industries in Western countries, and milk fats represent the most important dietary lipids in countries such as New Zealand. Milk fats perform many roles in bakery products, including dough strengthening, textural softeners, filling fats, coating lipids, laminating fats, and flavor improvers. This review reports how milk fats interact with the ingredients of main bakery products. It also elaborates on recent studies on how to modulate the quality and digestibility of baked goods by designing a new type of fat mimetic, in order to make calorie- and saturated fat-reduced bakery products. It provides a quick reference for both retailers and industrial manufacturers of milk fat-based bakery products.

Keywords: milk lipids; bakery products; fat replacer; shortening; baking activity

---

1. Introduction

Milk contributes approximately one third of human dietary lipid intake [1]. Milk lipids consist of protein and also neutral lipids (triacylglycerols(TAG), monoacylglycerols (MAG), diacylglycerols (DAG), free fatty acids (FFA)) and polar lipids (phospholipids) [2,3]. Milk fats and related dairy products, such as butter, anhydrous milk fats (AMF), cream, cultured milk fats, and cheese (matrix of milk lipids and proteins), have been incorporated into both extruded and baked products, including breads, cakes and biscuits [4,5].

There are several reviews on bread lipids functionalities [6], bakery fat replacers [7], bakery lipids [8], lipid shortenings [9], bakery emulsifiers [10], bread functional ingredients and textural improvers [11], milk lipids in the food system [12], and bread emulsifiers [13]. However, thus far, there has been no review on how milk fats perform their functions in bakery products. Therefore, this review aims to summarize milk fat applications in the bakery industry, and to update results on using milk fats to enhance the quality and nutritional value of baked goods. It also reports on the recent trends in relation to the health concerns of milk fats in baked products, and new ideas to reduce bakery energy density and saturated fatty acids (SFA).

2. Structure, Composition and Occurrence

2.1. Molecular Structure, Composition and Occurrence

Bovine milk lipids are comprised of 97.5% TAG, 0.36% DAG, 0.027% MAG, 0.027% FFA, and 0.6% phospholipids [14]. There are also some minor lipid classes present in milk, for instance, sterols, carotenoids, lipophilic vitamins, and flavor compounds [2].

The triacylglycerol molecule consists of a glycerol backbone and three fatty acids esterified at the positions of *sn*-1, *sn*-2, and *sn*-3. Two subclasses of phospholipids are glycerophospholipids and sphingolipids. Glycerophospholipids consist of a glycerol moiety with two fatty acids esterified at the positions of *sn*-1 and *sn*-2 and a hydroxyl group at *sn*-3 position, linked to a phosphate group and a hydrophilic residue. The structural details of the hydrophilic residue determine the types of glycerophospholipids, namely phosphatidylcholine (PC), phosphatidylserine (PS), phosphatidylethanolamine (PE), phosphatidylinositol (PI), phosphatidyl-glycerol (PG), and phosphatidic acid (PA) [3]. Sphingolipid consists of sphingosine backbone (ceramide, 2-amino-4-octadecene-1,3-diol), linked to a fatty acid through an amide bond and a polar head. Sphingomyelin (SM) is the predominant subclass of sphingolipids, having a phosphocholine head group. A minor constituent of sphingolipids in milk is glycosphingolipid, of which the polar group is comprised of carbohydrate groups (glucose, galactcose, and lactose) [15].

In intact raw bovine milk, lipids (3.3–4.6% [2]) are present in the form of milk fat globules (MFG), with an average diameter of 0.1–20 μm and are enveloped by a tri-layered phospholipid membrane [16]. The triple-layer membrane consists of a surface-active inner monolayer enveloping TAG in the center and an outer bilayer in contact with the aqueous phase of the milk. The milk fat globule membrane (MFGM) is composed of polar lipids, proteins, glycoproteins, enzymes and minor neutral lipids [17].

2.2. Fatty Acid Profile

The most abundant milk fatty acids are palmitic (26.3–30.4%), oleic (28.7–29.8%), stearic (10.1–14.6%), and myristic (8.7–7.9%) acids [14]. Anhydrous milk fats (AMF), known by the US Department of Agriculture (USDA) as 1003, consist of palmitic acid (27.7%), oleic acid (26.5%), stearic acid (12.8%), and myristic acid (10.6%) [2]. Due to a high content of stearic and palmitic fatty acids (melting points at 69.3 °C and 62.9 °C, respectively), milk fats are solid at ambient temperature. Conjugated linoleic acids (CLAs) are isomers of linoleic acids (0.8–2.5%) with the predominant CLAs being *cis*-9 and *trans*-11 CLAs (73–94%) [14].

SFA and monounsaturated fatty acids account for 62.2% and 28.9% (w/w) of the total fatty acids (FA) in the anhydrous butter oil of United States Department of Agriculture (USDA 1003), respectively, whereas long-chain FAs (LCFAs, C13–C21) accounts for 83.9% of the total FA, compared to medium-chain FA (MCFAs, C6–C12, 8.8%) and short-chain FA (SCFAs, C2–C5, 3.4%) [2]. Unlike LCFAs, SCFAs and MCFAs are absorbed intact as non-esterified fatty acids into the portal bloodstream and metabolized rapidly in the liver [18]. Via gastrointestinal digestion, medium-chain TAG is decomposed into glycerol and MCFAs, which reduces total cholesterol in serum by boosting hepatic synthesis of bile acid [19]. The SFA degrees of main shortening lipids are shown in Table 1 [2]. Lipids of dairy products can be separated by the Folch extraction [20], the Bligh method [21], the Röse–Gottlieb extraction [22], or dichloromethane [23]. Total lipid (TL) content of samples may be measured using gravimetric determination, a Gerber–van Gulik butyrometer, infrared spectrometry in a Milkoscan FT2 apparatus [22], or gas chromatography [24].

Table 1. Composition of shortening lipids.

USDA Code	Shortenings	Total Lipids (g)	SFA (g)	MUFA (g)	PUFA (g)	TFA (g)	SFA:UFA
4582	Canola oil	100.00	7.37	63.28	28.14	0.40	0.08
4506	Sunflower oil	100.00	10.30	19.50	65.70	-	0.12
4669	Soybean oil	100.00	15.25	22.73	57.33	0.68	0.19
4585	Margarine	80.32	14.20	30.29	24.17	14.95	0.26
4037	Rice bran oil	100.00	19.70	39.30	35.00	-	0.27
4615	Composite shortening	99.97	24.98	41.19	28.10	13.16	0.36
4002	Lard	100.00	39.20	45.10	11.20	-	0.70
1056	Cultured sour cream	19.35	10.14	4.59	0.80	0.80	1.88
1145	Butter	81.11	50.49	23.43	3.01	-	1.91
1003	Anhydrous butter oil	99.48	61.92	28.73	3.69	-	1.91
1017	Cheese cream	34.44	20.21	8.91	1.48	1.17	1.95
4513	Palm kernel oil	100.00	81.50	11.40	1.60	-	6.27
4663	Hydrogenated palm kernel oil (filling fat)	100.00	88.21	5.71	-	4.66	15.46
4701	Fully hydrogenated soy oil	100.00	93.97	1.34	0.38	1.15	54.50

Notes: Saturated fatty acids (SFA), mono-, poly-unsaturated fatty acids (MUFA/PUFA), and trans- fatty acids (TFA) of shortening lipids per 100 g adapted from US Department of Agriculture (USDA) Database v.3.9.5.3 [2].

2.3. Melting Properties and Solid Fat Index (SFI)

The SFI profile of milk fat crystal powder can be measured by pulsed nuclear magnetic resonance (p-NMR) with thermostatic incubation, and differential scanning calorimetry (DSC) can be used to determine the fat melting point [4]. The SFIs of major lipids in bakery products are illustrated in Figure 1 [25]. The SFI profile of milk butter is very similar to that of general use margarine, all-purpose shortening, and cake lipids, and thus, milk butter is interchangeable with other shortenings. Cocoa butter can be used for coating bakery products, whereas milk fats are too soft for coating. Even as a cookie filler, milk fats are not firm enough and need to be formulated with other lipids. To achieve optimum bakery activity, bakery lipids should have 20% SFI at 25 °C and a minimum of 5% SFI at 40 °C [26]. For instance, a blend of stearin fraction of palm-based DAG and palm mid-fraction (50:50 w/w; SFI at 30% and 10% for 25 °C and 40 °C, respectively; polymorphic form $\beta' + \beta$; slip melting point 55.4 ± 0.12 °C) makes a better bakery shortening than sunflower oil and palm oil [26]. An SFI profile of less than 15–20% at the dough temperature is too soft to make a shortening. However, fats that are too hard produce adverse effects, for instance, shortening with an SFI of ca. 47.5% at 20 °C produces less acceptable biscuits than shortening with an SFI of ca. 22.5% at 20 °C [27].

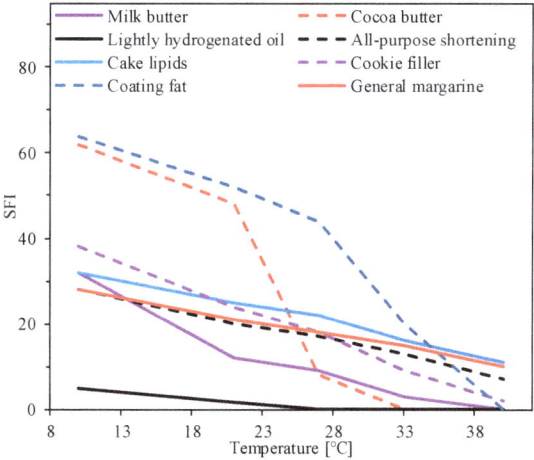

Figure 1. Solid fat index (SFI) profile of typical shortenings for bakery products. Notes: The SFI data was adapted from [25].

2.4. Crystalline Polymorphism

Aside from SFI, polymorphic forms of milk fat crystals are also a factor in controlling the bakery activity of lipids [4] Lipids exist in three major polymorphic forms, α, β′, and β, and their stability ascends in the order of α < β′ < β. Lipid crystal α is usually undesirable due to its instability. The β crystal (large plate-like) is stable, but coarse and sandy, whereas β′ form is desired in baked goods since it is fine, needle-shaped, and stable. Enzymatic inter-esterification can rearrange fatty acids on the TAG backbone, creating tightly packed, small β′ crystals, which produce more desirable bakery activity than composite blends [28]. Milk fats, together with other natural edible lipids such as tallow, palm oil, cottonseed oil, and high erucic acid rapeseed oil possess the β′ polymorph, whereas crystals of soybean oil, sunflower oil, coconut oil, palm kernel oil, and lard are usually present in the β polymorph. Crystalline forms may transform into a more stable form as time and temperature change [29]. The crystalline polymorphism can be determined by using an x-ray diffractometer (XRD) [30]. Characteristic peaks of the acylglycerol emulsifier-shortening blend at 4.15 Å and 4.6 Å are from α and β forms, respectively, whereas the β′ form demonstrates three signals at 3.8 Å, 4.2 Å and 4.3 Å. In addition, sub-β and sub-β′ forms may cause peaks at 4.5 Å and 4.0 Å, respectively [31].

3. Milk Fats for Bakery Products

3.1. Milk Fats and Related Dairy Products

Milk fats and related dairy products include butter, anhydrous milk fat, ghee, and cheese (combination of milk lipids and proteins) [32]. Milk butter is the predominant milk fat product used in the bakery industry, comprised of 81.11% milk lipids and 16.17% moisture, approximately (USDA 1145 in Table 2 [2]). In native milk, the enveloped fat globules are dispersed in the serum [33]. During churning, the membrane is disrupted and those milk fat globules aggregate to form butter, separating out from serum (buttermilk) [34]. Cultured lactic butter is more popular in Europe than in the USA, whereas sweet cream butter is more prevalent in the UK and USA than in other countries [35]. Salted butter (1.6–1.7% salt) has a 4-fold shelf life in refrigeration than unsalted butter due to reduced water activity [32]. Ghee is clarified milk fats from butter or cream, with an enriched flavor [36]. A milk butter blend with vegetable oil (e.g., corn oil, canola oil) reduces the overall SFA. Being hydrogenated from vegetable oils (e.g., soybean and palm oil) or animal fats (e.g., beef tallow) to raise the SFI and melting point, margarine is a cheaper substitute for milk butter in the bakery industry [37]. To avoid TFAs resulted from hydrogenation, inter-esterification of vegetable oils (soybean oil, palm stearin, coconut stearin; 20:50:30, $w/w/w$) by Lipozyme RM IM produces optimized SFI and crystals of β′ polymorphic form, which are equally as effective as commercial margarines [38].

Table 2. Proximate nutritional information of main milk lipid products.

USDA Code	Milk Fats	Water (g)	Energy (kJ)	Protein (g)	Lipids (g)	Ash (g)	Carbohydrate (g)
1017	Cheese	52.62	1466	6.15	34.44	1.27	5.52
1005	Cheese	41.11	1553	23.24	29.68	1.85	2.79
1009	Cheddar cheese	36.37	1684	22.87	33.31	3.71	3.37
1053	Cream	57.71	1424	2.84	36.08	0.53	2.84
1056	Cultured sour cream	73.07	830	2.44	19.35	0.51	4.63
1145	Butter	16.17	2999	0.85	81.11	0.09	0.06
1003	Anhydrous butter oil	0.24	3665	0.28	99.48	-	-

Notes: The nutritional data was adapted from US Department of Agriculture (USDA) Database v.3.9.5.3 [2].

AMF contains 99.48% lipids and 0.24% moisture, respectively (Table 2 [2]). According to the Codex/CFR Alimentary, AMF and butter oil must be comprised of no less than 99.8% and 99.6% lipids, respectively, without additives [32]. The AMF is produced by vacuum drying and removal of nonfat solids from pasteurized cream. First, cream (40% lipids) is concentrated to 70–80% milk fats, and after phase inversion, the milk fats are further dried to no more than 0.1% moisture [32]. The AMF can be produced from both butter and cream [33], and butter oil is made out of butter [32]. For cost-saving, substitution of 30% AMF by hydrogenated vegetable oils (e.g., soya or coconut oil) has been formulated into the shortening of bakery products in Asian countries such as Japan [25]. AMF has a broad melting and crystallization range, fully crystallizing at −40 °C and completely melting at 38–40 °C. Thus AMF can be fractionated into low (<10 °C), middle (10–20 °C), high (>20 °C), and very high (>50 °C for confectionery) melting fractions [39].

Cheese is produced from milk by inoculation with bacteria and separation of the resulting semi-solid curd (33–55% lipids for origin cream cheese, 0.5–16.5% lipids for reduced-fat cheese) from the liquid whey, leading to less-perishable products than milk [40]. Among the most commonly used bakery flavors, cheddar and parmesan cheeses have been used to impart flavor in biscuits or crackers [41]. Aside from flavor enrichment, cheese can also be used as a bakery coating or filling lipids [42]. A typical cheddar cheese contains 33.31% lipids (Table 2 [2]). Gas chromatography analysis of enzymatically-modified white cheese for bakery flavor revealed 58 volatile compounds of seven chemical classes including alcohols (12), aldehydes (8), ketones (10), esters (8), acids (11) and hydrocarbons (9), among which most compounds were produced by metabolism of carbohydrates, milk fats and amino acids [42]. Kefir cheese culture fermentation yielded volatile compounds, for instance, diacetyl (i.e., major buttery aroma), acetaldehyde, ethanol, and acetone by metabolism of probiotic bacteria (e.g., lactic acid bacteria, *L. acidophilus*, *Bifidobacterium* spp.) and yeasts (e.g., *Saccharomyces* spp. and *Kluyveromyces* spp.) [43]. Bacterial metabolism produces diacetyl, for instance, by *Lactococcus lactis* subsp. [44].

Sour cream, a critical bakery flavor improver, is produced by the moderate-temperature fermentation of cream, and it can also be made by the treatment of acid-producing bacterial cultures on pasteurized cream. Compared to cream, sour cream (typical lipid content 19.35% in Table 2 [2]) is thicker and more acidic, with a longer shelf-life [45].

Furthermore, milk fats are often consumed together with biscuits and breads (e.g., as fillers) [46]. Milk fat products such as butter and AMF are sometimes manufactured as flaked or powdered forms by spray chilling or spray drying, which are easy to disperse [29]. The typical composition of milk fats and related dairy products are illustrated in Table 2 [2].

3.2. Functional Roles of Milk Fats in Bakery Products: Baking Activity

Milk fats have been used to perform multifunctional roles in bakery products, for instance, as mouthfeel and flavor improvers, texture improvers, dough conditioners, and anti-staling agents [29]. In addition, milk fats can fulfil a wide variety of functions such as laminating and filling fats, coating or topping lipids, spray oil, and imparting flavor [35]. The functions of milk fats are dependent on the dose and the type of baked products. For instance, they play more strengthening roles in yeast-leavened bread dough than in cookie/biscuit dough or cake batter, whereas cake fats are highly attributed for aeration and whipping in batter agitation [47], and biscuit or cookie laminating fats are mainly responsible for crisping and puffy effects by textural improvement [47]. Almost half of the lipids in coconut oil (USDA 4047) are comprised of lauric acid (41.84%, T_m = 43.2 °C), but bovine milk fats (USDA 1003) contain only 2.79% lauric acid in comparison [2]. Therefore, milk butter needs to be blended with cocoa or equivalent to make a bakery coating.

Bakery shortening is defined as the ability of a fat to lubricate, weaken, or shorten the structure of bakery products, thereby providing tenderization effects and other desirable textural properties to bakery products [48]. During the mixing process of dough or batter, lipids interact with gluten and starch particles to strengthen their network, thus improve the gas retention of dough. Hence, bakery products become softened, resulting in consistent grain, lubricated mouthfeel, enhanced heat transfer, and extended shelf life [9]. Shortening lipids are made from milk butter, animal fats (e.g., tallow, lard), or hydrogenated plant oils (e.g., palm oil) [47]. In contrast to standard shortening, lipids such as hydrogenated vegetable fats may be used to replace milk fats for bakery products, such as biscuits [49].

Laminated dough shortening has an SFI of 10–40% for the temperature range of 33 °C to 10 °C, causing a puffy texture for croissants, danishes, and pastries. Milk butter is a benchmark laminated dough preparation agent for appropriate SFI profile and β'-form crystal. Cheap alternatives include hydrogenated shortenings and inter-esterified fats, which lead to a trans fatty acids (TFA) issue or less acceptable sensory quality [4].

Bakery lipids have their characteristic SFI profile, plasticity (processability), and antioxidant stability [50]. For instance, a coconut oil cookie filler is designed as a 59% SFI at 10 °C, 29% at 21.1 °C, and 0% at 26.7 °C onwards, with a melting point of 24.5 °C. In contrast, croissant shortening melts at 39 °C, with an SFI profile of 39% at 10 °C, 27% at 21.1 °C, 22% at 26.7 °C, 19% at 33.3 °C, and 18% at 43.3 °C [47]. Milk fat flavors have been attributed to volatile molecules, including branched-chain fatty acids, lactones, methyl ketones, aldehydes, and other minor compounds, which are originated from milk fats or produced during fermentation, lipolysis, or processing. Milk fat products, such as cheese (e.g., cheddar and feta), cream, sour cream, and butter are all used to improve the sensory properties of bakery products [43].

3.3. Interaction of Milk Fats with Other Bakery Ingredients

3.3.1. Lipid–Protein/Starch Interaction

Lipid–protein binding interactions can increase gluten polymerization. However, the ionic amphiphilic binding will cause interface aggregation due to charge neutralization, and therefore, this interaction may also decrease surface activity as the lipid concentration at the aqueous–oil interface increases to a certain level, which leads to the disruption of protein–protein interactions in the interfacial film [51]. The gluten–lipid interaction yields a dynamic balance of surface activity, altering the surface activity and aeration ability. This mechanism is critical for dough rheological characteristics and product textural properties. Horra et al. [52] compared refined milk fats (SFI 38% at 25 °C) and margarine shortening (SFI 5–25% at 25 °C) and found, through confocal microscopy, that the gluten network with milk fats is less developed and more orderly structured (with isolated starch particles) than the network formed with margarine shortening, thereby producing greater elastic and viscous moduli, and higher puff pastry.

During dough mixing, milk fats coat the gluten network and starch particles, reducing the water hydration capacity of the dough [6]. With the formation of an extensible gluten film by hybrid hydration and lipid coating, the lipid crystals decrease the surface tension of the gluten film (lubrication effects), promoting aeration of the dough [9]. The crystals align their orientation along the air cells and stabilize them. Milk fats (β' polymorph) aerate more effectively than soybean oil (β polymorph) by forming fine and consistent gas bubbles [53].

During dough fermentation and proofing, the fat crystals further melt and become absorbed at the gas–liquid interface with elevated temperatures. They re-orientate along the interface plane and hold the yeast-leavened carbon dioxide in the gas cells [54]. Low melting fats or oils have been found to be much less effective in gas retention at this stage [55].

During baking, starch particles become gelatinized and the gluten film turns into a permanent cross-linked thin film together with lipids, with the crust drying and browning (the Maillard reaction) taking place concomitantly [56]. Without lipids, the bubbles tend to coalesce or collapse and produce coarse crumb grain, whereas shortening fats lead to fine crumb grain and consistent porosity [57]. Hydrogenated fats produce a stronger dough and more tender cookies than sunflower oil [58].

When the baked products cool down, amylose and amylopectin crystallize and retrograde in the early stages of storage and over the course of shelf life, respectively [59]. Using low-frequency NMR, it has been found that water migrates from crumb to crust or to amylopectin during storage, and immobilization of moisture will reduce water activity and decrease crystallization of amylopectin, thereby inhibiting the staling rate [60]. Most bread staling mechanisms can be explained by water migration and therefore reducing water activity by changing the gluten network and reducing hydrophilicity of the bread crumb can slow down the rate of bread crumb staling. During bread storage, starch polymer retrogrades concomitantly with fat re-crystallization [61]. This concurrent polymorphic conversion from β′ to β has been evidenced by powder XRD analysis for croissant samples [62].

In brief, shortenings such as milk fats have effects on the lubrication/stabilization (mixing), gas retention (proofing), textural tenderization (baking), and anti-staling (storage) properties of bakery products. The addition of emulsifiers can consolidate the above effects, thus reducing the amount of shortening lipids required. The interaction of lipids with other ingredients is different among breads, cookies, biscuits and cakes, such as with the additional interaction of lipids with egg components.

3.3.2. Starch–Lipid Complexes

Starch can form complexes effectively with MAG, as it can with fatty acids, but TAG does not form complexes with starch [63]. A previous report has shown that four kinds of lipids: monopalmitate glycerol (96.3%), (palmitic acid (41.8%), dipalmitate glycerol (DPG, 1.1%), and tripalmitate glycerol (8.3%) have reduced complexing ability [63]. The starch–lipid complexes have been found to lower the glycemic load of bakery products and impact on their staling processing. Using confocal laser scanning microscope (CLSM) and scanning electron microscopy (SEM), both non-inclusion and inclusion lotus seed starch–lipid complexes have been identified, causing slow digestibility of starch [64]. The complexing index of debranched starch–stearic acid complexes reached 89.31% [65], while that of the native starch (yam)–palmitic acid complexes (2%, w/w, starch base) was maximized as 26.39% [66]. In addition to the reduction of the starch glycemic index, high amylose corn starch–lipid complexes inhibited the staling process of baked goods [67], as also evidenced by a recent report, where the firmness of wheat bread during storage was significantly reduced by resistant starch [68].

3.3.3. Emulsifier Functionalities during Dough/Batter Forming

Emulsifiers can be used to disperse milk fats, enhance their baking activity, assist ingredient blending and emulsification during dough/batter formation, and promote aeration and air distribution, especially for cake batter [29]. Commonly used emulsifiers for baked goods include MAG and DAG (E471), lecithin (E322), sodium stearoyl lactylate (SSL, E481), and diacetyl tartaric acid ester of mono- and diacylglycerols (DATEM, E472e) [11]. For instance, in a high-ratio layer cake, 5% MAG was formulated into the shortening [31].

Similar to the baking activity of shortening, anionic emulsifiers such as DATEM, SSL, and calcium stearoyl-2-lactylate (CSL) are useful in both dough strengthening and bread softening, as are the nonionic emulsifiers (sucrose esters of fatty acids (SE), polysorbate-60 (poly-60)). Lecithin and distilled MAG have no strengthening effects [13]. Fu et al. [31] compared distilled MAG and four acylglycerols (40%) of octanoic acid (8:0), palmitic acid (16:0), stearic acid (18:0), and linoleic acid (18:2) and found that monopalmitate glycerol and monostearin glycerol led to a higher SFI and finer crystals (β′ form), thus increasing aeration ability in batter formation and tenderizing the crumb of layer cakes. In contrast, monooctanoic glycerol and linoleic acid glycerol produced adverse effects to the SFI and β′ form crystals, thereby reducing cake size and increasing its firmness. In addition, lecithin and distilled monostearate stabilized the shortening crystals and increased the air-absorbing ability on both beef tallow and hydrogenated palm oil [69]. Using digital imaging of crumb micro-structure, the emulsifier functionality in assisting air aeration was recognized. At the same level of dough hydration, five emulsifiers (DATEM, SSL, distilled MAG, lecithin, and polyglycerol esters of fatty acids (PGEF)) increased the bread dough permeability and gas retention ability, resulting in increased gas bubble number and homogeneity [70].

4. Milk Fats for Bakery Products

Breads use less fats and sugar than biscuits and cakes, and biscuit recipes use less water than breads and cakes. For instance, breads (AACC 10–10 recipe, flour based) are comprised of wheat flour, 6% sugar, 5% milk butter, 1.5% salt, 1.5% yeast, and 60% water; biscuits (AACC 10–54) consist of wheat flour; 42% sugar, 40% shortening, 1% skim milk powder, 1.25% salt, 1% sodium bicarbonate,

0.5% ammonium bicarbonate, 1.5% high fructose corn syrup, 22% water; high-ratio cakes (AACC 10–90) are made out of wheat flour, 140% sugar, 50% shortening, 2% emulsifier, 12% dry skimmed milk powder, 5.5% baking powder, 9% egg white powder, 3% salt, and ca. 135% water [71]. Cakes are distinct from other products for containing egg, though some other products may also contain egg.

4.1. Bread Fats

Milk fats account for 3–4% of a bread recipe [71]. During dough mixing, both starch particles and gluten become hydrated, and the gluten proteins polymerize through reactions between the sulfhydryl (–SH) groups and disulfide (–SS–) bonds, forming an extensive, interlinked dough skeleton [72]. Milk fats mainly perform three kinds of functions in bread dough. First, lipid crystals brace the developed gluten network as a plasticizer. In this instance, shortening oil (e.g., sunflower oil) exhibits far less of an effect on the developing strong gluten network than shortening fats and milk butter due to less SFI [58]. Secondly, lipid crystals align themselves to the gas–liquid interface of bubbles during dough mixing, exerting lubricating effects [73]. Lastly, the lipid crystals enhance the stability index of bread dough [74], and the β′ crystal-stabilized bubbles are larger than that of β crystals [75]. β′ crystal lipids aerate dough more effectively than β crystal lipids [76].

During dough fermentation at 40 °C, yeast digests glucose and emits carbon dioxide and ethanol. Newly produced carbon dioxide diffuses into gas bubbles and leavens dough to 1–1.5-fold in height [77]. Milk fats will then melt and form an extensible thin film, further stabilizing the gas bubbles [78], whereas doughs with insufficient lipids will leak gas via the gluten network due to the penetration or rupturing of the cell wall [78]. However, a high concentration of lipids will inhibit dough rising, as aggregated gluten and solid lipid crystals exert low elasticity, thereby hindering the expansion of bubbles [63].

Upon heating, the cells expand with carbon dioxide diffusion and moisture/ethanol evaporation. With the moisture mobilization and heat transfer, gluten and gelatinized starch become solidified and form a fine crumb texture, while at the same time, the bread crust dries and turns brown due to the Maillard reaction [73]. Shortening fats melt fully and form an elastic thin film together with gluten along cell walls, again stabilizing gas cells [54]. Solid fat-incorporated breads exhibit increased porosity, loaf volume, and softness [79]. During baking, with the melting and gelatinization of starch particles, fat globules melt and form gas cells. The dough moisture migrates towards the edges of gas bubbles to evaporate. Eventually, the bread forms an interlinked porous texture [80].

During the staling process, the bread crust becomes leathery and the crumb turns rigid and unresilient, in parallel to the losses of aroma and eating quality [81]. The migration of moisture across the crumb and crust leads to elevated bread rigidity. In addition, amylose and amylopectin retrograde successively over shelf life [56].

There are several approaches to delay the staling of bread, for instance, by the addition of plasticizers, cross-linkers and fillers, or by the modulation storage temperature to inhibit deformation during staling [61]. Milk fats have sufficient SFI at ambient temperature, and thus, they are able to act as plasticizers to increase storage stability, as well as change the thermoplastic properties.

4.2. Biscuit Fats

Biscuits are among the most consumed bakery products worldwide, and they are formulated with flour, fat, sugar, milk, water, eggs (optional), and salt into a viscous dough, and are baked on a flat surface [71]. In addition to lubrication and aeration in dough forming, biscuit fats perform such roles including filling, laminating, coating, surface spray, nutritional value, sensory, and tenderization. Fats (T_m ca. 33 °C to give smooth mouthfeel, SFI 53% at 20 °C and 3% at 35 °C) constitute around half of the biscuit filler, in which inappropriate melting points will cause brittleness or filling collapse. Coating fats are usually cocoa butter equivalents (T_m ca. 36.6 °C), whereas typical spray lipids approximately possess SFI profiles of 22% at 20 °C and 0.5% at 35 °C [29]. Milk fats can be formulated compositely to fulfill these roles. Milk butter (no less than 7%, flour based) and cheese have been used to make premium butter biscuits (USDA 18214 [2]) and cheese crackers (USDA 45080543 [2]), imparting a buttery aroma. Moreover, cookies also utilize milk butter powder to laminate the dough sheets into several discrete layers, creating a puffy effect on the end product [29]. Enzymatically hydrolyzed or cultured milk fats have been used as flavor agents [42,44]. Aside from the above functionalities, milk butter also serves as a nutritional ingredient. For instance, AMF and butter is comprised of polyunsaturated fats and lipophilic compounds such as vitamin E and β-carotene [35]. In addition, milk fats are natural lipids, without the trans-fatty acid issues of other vegetables fats such as hydrogenated shortenings [82].

Crackers are usually salty biscuits, based on layered dough, whereas cookies are normally made out of high fat and sugar recipes (short-dough [83], more cake-like). To counterbalance gluten development with syrup, comparable fats are added to confine starch granule swelling and limit dough forming [29]. Cookie dough is short-formed, and therefore a chemical leavening agent is used to increase its volume. The lipid content of leavened cookies and crackers is 7–20%, whereas unleavened cookies can have a lipid content as high as 16–33% (dough-based, Table 3). A typical cracker recipe incorporates 23.1% of milk butter (flour-based) [35]. In contrast to breads (35–45% moisture), the moisture content of cookies and biscuits are comparably low. For instance, crackers and cookies in contain 2.75% and 5.9% moisture, respectively [27], and thus they can sustain a long shelf life. Compared to cookies, cracker recipes have no sugar (Table 3).

Table 3. Biscuit recipes based on 100 flour.

Ingredients (g)	Cracker 1	Biscuit 1	Biscuit 2	Biscuit 3	Biscuit 4
Wheat flour	100.00	100.00	100.00	100.00	100.00
Water	27.50	35.71	13.33	20.00	20.00
Shortening	10.50	13.84	44.89	39.90	66.00
Baking Powder	0.80	0.98	1.11	0.50	-
Salt	1.00	0.66	0.93	0.71	2.40
Emulsifier	2.75	0.59	5.00	0.51	1.00
Sugar	-	26.79	60.00	40.40	33.00
Shortening Dough Base	7%	8%	20%	20%	33%
Reference	[84]	[85]	[86]	[83]	[87]

4.3. Milk Fats in Cakes

Cake batter is an emulsion of flour, sugar, shortening, egg, and other minor ingredients [88], and cakes contain more lipids and sugar than breads. Compared to biscuits, milk fats, especially butter, play a greater role in cakes than in biscuits. Yellow cakes (Table 4) use butter and whole egg, resulting in a rich color, tender grain, and milky flavor. White cakes, on the other hand, usually use egg white and shortening instead of milk butter (Table 4). Pound cakes require equal amounts of flour, whole egg, milk butter, and sugar (Table 4), leavened by baking powder. Distinctly, butter may be absent in sponge cake recipes, where egg performs the aeration function in batters, creating foam and an airy grain.

In a layer cake formula ([89], Table 4), the amount of sugar is not greater than the quantity of wheat flour (both 100 g), and the egg amount (49.1 g) is equal or comparable to the amount of shortening (40.91 g). The amount of liquid in the recipe (114.89 g, Table 4) may be equal to or greater than the amount of sugar in the recipe (100 g, Table 4) [90]. Conversely, high-ratio cakes use more sugar than wheat flour (Table 4). To counterbalance the inhibitory effect of excess sugar on starch gelatinization [90], an extra amount of egg is added to strengthen the formula (60 g egg vs. 40 g shortening [31] in Table 4).

Cake fats perform similar functions to bread lipids, for example, air incorporation, air cell stabilization, structure tenderization, and elevation of oven spring [91]. Propylene glycol monostearate (PGMS, 1.8% w/w), glycerol monostearate (GMS, 1% w/w), and lecithin (0.8% w/w) blended with soy bean oil are equally as effective as commercial liquid shortenings in increasing cake size and softness.

However, liquid shortening cakes exhibit a reduced firming rate compared to cakes containing plastic shortening, seen over the course of three-weeks in storage [91].

Table 4. Cake batter recipes based on 100 g wheat flour.

Ingredients (g)	White Cake *	Yellow Layer Cake	High Ratio Cake	Pound Cake	Sponge Cake	Sponge Cake
Wheat Flour	100.00	100.00	100.00	100.00	100.00	100.00
Granulated Sugar	136.00	100.00	120.00	100.00	81.82	100.00
Water	106.00	65.79	75.00	16.33	-	-
Fresh Egg	60.00	49.10	60.00	100.00	127.27	100.00
Butter or Equivalent	25.00	40.91	40.00	83.55	86.36	100.00
Emulsifier	0.30	-	2.00	0.12	-	-
Skimmed Milk Powder	9.00	8.18	7.00	-	6.05	6.66
Vanilla Flavor	-	2.03	-	-	-	-
Salt	3.00	2.03	3.00	1.63	-	-
Baking Powder	6.00	0.55	5.50	6.53	9.09	-
Fats in Batter	6%	11%	10%	20%	19%	25%
References	[35]	[89]	[31]	[92]	[93]	[94]

Notes: * White cakes use egg white (not yolk) and shortening, instead of butter.

5. Milk Fats Replacement

To make cost effective bakery products, these milk fats need to be replaced with more economic sources. In addition, milk fats are high-calorie (3665 kJ/100 g for AMF in Table 2), highly-saturated lipids (ca. 66% in Table 1). Table 5 illustrates some fat replacers used in bakery products. The high content of shortenings in biscuits and cakes, together with the total lipid content in some bakery products (1812 kJ/100 g in Table 6), catalogues them as high-calorie foods (>1675 kJ/100 g [95]), as shown in Table 6. In order to produce bakery goods that are low in calories and saturated lipids, milk fats need to be substituted. There has been interest in using resistant starch (RS) emulsions to substitute bakery fats by 25–50%. In this regard, four forms of starch (2.6–46% RS) exhibited great potential in improving cookie/cake size and symmetry due to the extra hydration capacity of the added starches, while maintaining color and sensory score [96]. Specialty fats (e.g., hydrogenated fats) in the bakery industry have been used to improve texture, shelf life and sensory acceptance. However, they are associated with high serum levels of low density lipoprotein and cholesterol, and the subsequent development of atherosclerosis [30]. Oleogels are recent alternatives to reduce SFA, as illustrated in Table 5. It has been found that fat replacement has less impact on the acceptability of biscuits than sugar reduction [97].

Table 5. Successful fat replacer in bakery products.

Figure	Bakery Products (Flour 100 g)	Replacement % of Full-Fat Bakery Products	Results	References
Beeswax–sunflower oil oleogels	AACC 10–90 cake	100%	SFA 58%→15.5%	[98]
Candelilla wax–canola oil oleogels	AACC 10–54 cookie	30–40%	SFA 63.4%→32.3%	[50]
Carnauba wax–canola oil oleogels	AACC 10–90 cake	25%	SFA 74.2%→64.24%	[99]
Candelilla wax–canola oil oleogels	AACC 10–52 cookie	100%	SFA 52.8%→8.5%	[100]
Inulin from chicory roots.	Sponge cake: 100% sugar, 46% sunflower oil	70%	Reduced fat and fortified fiber	[101]
Inulin	Short dough biscuit: 74.1% margarine; 37% sugar	25%	Textural and sensory properties maintained	[102]
Inulin	Short dough biscuit: 30% shortening; 15% sugar	20%	Weakened lubrication of biscuit dough	[103]

Table 5. Cont.

Figure	Bakery Products (Flour 100 g)	Replacement % of Full-Fat Bakery Products	Results	References
Acetylated rice starch	Cookie: 60% sugar, 30% shortening	20%	Native and modified rice starch equally effective	[104]
Inulin	Biscuit: 45% margarine, 26.7% sugar	20%	Biscuit energy density reduced by 580 kJ/kg	[105]
Corn fiber, maltodextrin or lupine extract	Short dough biscuit: 132% margarine, 66% sugar	30%	28.6% fat reduction and 23 g/kg fiber fortification	[106]
Carnauba wax 5%—cotton oil oleogels	AACC 10–90 cake	50%	SFI similar to shortening fats	[107]
Chia seeds mucilage (80.16% carbohydrate, 10.63–10.76% protein)	AACC 10–90 cake and AACC 10–10 bread	50% and 75%	51.6–56.6% fat reduction and protein fortification	[73]
High-oleic sunflower oil and inulin/β-glucan/lecithin	Biscuit: 34% sugar, 46% shortening	100%	Lecithin (3%, sunflower based) achieved similar sensory quality	[108]
Chia mucilage gel	Pound cake: 75% sugar, 30% shortening	25%	Higher replacement led to adverse effect to color and texture	[109]
Puree of canned green peas	full-fat chocolate bar cookies: 324% sugar, 134% shortening	75%	By sensory assessment	[110]
High oleic sunflower oil + wheat bran (1.9:1)	Cookie: 52% sugar, 33% shortening, 40% egg	100%	SFA: 54.6→24.5%	[111]
77.3:34:12.4:1.3 olive oil: water: inulin: lecithin	Cake: milk fats 35%, sugar 33%, egg 40%	50%	SFA: <39% TL: <19%	[112]

Notes: Solid fat index (SFI); saturated fatty acids (SFA); total lipids (TL).

In contrast to the moisture-retention and staling-retardation effects of carbohydrate-based replacers, protein-based replacers perform functions as texturizers. For instance, milk whey protein concentrate has been compositely used to substitute fats [113]. In addition, enzymes can also reduce shortening use, by targeting the endogenous flour lipids. Fugal lipase, e.g., Lipopan F, has been successfully developed to hydrolyze flour lipids to replace milk fats [79,114,115]. In another report, amylase-hydrolyzed starch was used to replace shortening, and achieved a comparable loaf size and consistency, but lower springiness and softness [116]. In general, reduced-fat bakery products have shown poorer performance in regards to mouthfeel, flavor, and texture properties than standard bakery products [117].

Table 6. Nutritional information of baked goods by US Department of Agriculture (USDA) [2].

Baked Products	Bread	Biscuit	Cookie	Sponge Cake	Pound Cake	Wheat Cracker	White Cake	Yellow Cake
USDA code	18064	21142	3213	18133	45209528	18232	45262644	45174254
Water (g)	35.25	27.88	5.90	29.70	26.25	2.75	24.69	18.83
Energy (kJ)	1145	1547	1812	1213	1516	1903	1653.865	1725.044
Protein (g)	10.67	7.08	11.80	5.4	3.75	7.3	2.47	2.35
Total lipids (g)	4.53	18.92	13.20	2.7	15.00	16.4	18.52	16.47
Ash (g)	2.01	3.31	2.00	1.2	-	2.83	-	-
Carbohydrate (g)	47.54	42.82	67.10	61	55.00	70.73	54.32	62.35
Sugar (g)	5.73	3.88	24.2	36.66	40.00	6.9	-	-
Total dietary fiber (g)	4.00	2.50	0.20	0.5	-	15.48	43.21	49.41
Total saturated FA (g)	0.70	11.80	2.35	0.80	3.75	3.21	7.41	8.24
Total monounsaturated FA (g)	0.61	2.49	5.99	0.95	0.00	3.47	-	-
Total polyunsaturated FA (g)	1.62	2.20	2.88	0.45	0.00	8.474	-	-
Total trans FA (g)	0.03	0.21	0.02	-	3.75	0.034	-	-

5.1. Carbohydrate-Based Milk Fat Mimetics

Carbohydrate-based fat mimetics are the most common milk fat replacers, including plant polysaccharides, dietary fiber, and starch [118]. These fat mimetics have been initially designed to generate sufficient baking activity, such as moisture retention, texturizing, and mouthfeel, whereas yielding only half to a quarter of the total calories of fats. However, in terms of flavor, palatability, crumb consistency, appearance, and customer acceptances, these replacers are less effective than milk fats. Recently, dietary fiber (e.g., inulin [119] and pectin [120]) and other resistant starches (e.g., Emjel) have been added to cookie and cake recipes [96], and they achieved similar textural properties to full-fat bakery products. Pectin (Yuja pomace) gel substitution (10%, w/w) led to the same level of volume and textural properties as shortening cake (AACC 10–90), with elevated softness and whiteness [120]. Inulin (e.g., *Agave angustifolia* fructans) replacement (20%) led to similar sensory properties and enhanced prebiotic activity [121]. Light microscopy images showed that, with the replacement of shortening fats with β-glucans from an edible mushroom in the batter recipe, the population of gas bubbles became decreased, with broader size distribution, which indicated the loss of stabilization by forming an interfacial lipid film along bubbles during batter forming [122].

5.2. Lipid-Based Milk Fat Mimetics

Unsaturated lipids or low-calorie lipids have been used to replace milk fats. For instance, replacement of butter in breads by rapeseeds caused a 91% reduction of low-density-lipoprotein-cholesterol in plasma [123]. Margarine is a cheap alternative to milk fats. However, the high water content of margarine limits its use in biscuit manufacturing. Animal fats have been used to inter-esterify with plant oils (e.g., canola oil) to prepare bread shortenings [124], and cookies prepared with oils were firmer than full-fat cookies [125], whereas shortening (palm oil) and emulsifiers together have produced cakes with a similar firmness to cakes prepared with fats [91]. Inter-esterified beef tallow caused slower crystallization than tallow, and brought about an SFI increment of approximately 11% and 5% at 25 °C and 40 °C, respectively, thus increasing cake size and textural consistency. Inter-esterification of the beef tallow-palm medium fraction produced similar plasticity and operability of shortening to beef tallow [126]. The addition of MAG and tripalmitin induced the formation of a polymorphic β-form, accelerating the processing of crystal formation and reducing the size of crystals [127].

5.3. Emulsion-Based Milk Fat Mimetics

Oleogels have been fabricated to structure vegetable oils for bakery products, in order to reduce SFA and trans-fatty acids from the diet, as illustrated in Table 5. Oleogels were fabricated by thermal dispersion of sunflower oil into SSL (7–13%, w/w) at 75 °C [30]. Candelilla wax–canola oil oleogels reduced cookie SFA to ca. 8%, without damaging eating quality [100]. In another study, beeswax–sunflower oil oleogels reduced SFA in cakes to 14–17% from 58% in full-fat cakes [98]. In a previous report, monoacylglycerol organogels and sunflower oil-loaded hydrogels were used to replace shortening fats (palm oil), by 81% [128]. MAG–sunflower/palm oil (0.5%/7%, flour based) water gels have been formulated into bread recipes (4.7% MAG, 55.8% oil, and 39.5% water, w/w/w) [129]. Edible oleogels enhanced nutritional profiles and bioactive benefits [130], and showed important features, such as thermo-reversibility and thixotropy [131]. SSL (7%) has been used as a gelling agent to structure sunflower oil oleogels, creating a crystal network similar to that of TAG [30].

Gels of hydroxypropyl methylcellulose (HPMC)/sunflower oil produced more acceptable biscuits than milk fat, vegetable shortening, sunflower oil/xanthan gum, olive oil/HPMC, and olive oil/xanthan gum [132]. A 15% replacement of HMPC/inulin made crisper biscuits than full-fat shortening [133]. Biscuit dough formulated with an HPMC emulsion showed similar rheological properties to dough made out of shortening fats [134].

5.4. Whole Foods or Combined Ingredients to Replace Bakery Lipids

Whole foods, such as avocado, chia, and banana, have been used to replace bakery lipids. For instance, chia (*Salvia hispanica L.* oil content 30–40%, protein content 15–25%) is comprised of rich polyunsaturated fatty acids, such as ω-3 fatty acids (linolenic acid, 54–67%) and ω-6 (linoleic acid, 12–21%). A chia mucilage gel (25%) has been shown to be a feasible alternative for pound cake shortening [109]. The use of oatrim (100%), bean puree (75%) or green pea puree (75%) as fat replacers in biscuits have proven to be equally effective, and avocado puree can replace half of the shortening in both cakes and biscuits [7]. Okra gum from an edible green fruit (flowing plant of the mallow family) has been identified as a fat replacer for reduced-calorie bakery products, improving the nutritional quality of baked goods [135]. Avocado purée as a full replacement of shortening fats has brought about an increase in MUFA by 16.51%. Substitution by half demonstrated comparable acceptability, whereas further fortification with avocado purée caused undesirable flavor and aftertaste, according to the tested panelists in the study [136].

6. Conclusions

This review verifies the relevance and significance of milk fats in bakery products. Their roles include altering structural, rheological, nutritional, and sensory characteristics. The milk fats can be used for dough strengthening in bread making, texture softeners in cakes, and sensory improvers in butter biscuits. In addition, they can be used as cookie fillers, laminating fats, topping and coating fats in bakery products. The interaction of milk fats with flour gluten and starch particles provides dough strengthening and texture improving effects to bakery products. Appropriate fat substitution with the design of new matrices such as oleogels and inulin gels can improve the nutritional value of bakery products by reducing the saturated fatty acid content and energy density, and by increasing the nutrient quality, without adversely affecting the textural and sensory properties. In addition, lipase treatment of flour lipids or milk fats can generate emulsifiers including monoacylglycerols, which may enhance the shortening effect of milk fats and thereby reduce shortening use. Milk fatty acid–wheat starch complexes may also be facilitated so as to reduce glycemic response and increase the shelf-life of baked goods.

In conclusion, milk fats have performed multi-functions in both technical importance and nutritional values, especially for high-end, valued-added baked goods. With partial replacement of milk fats in bakery products to balance their saturated lipids, both nutritional quality and customer acceptability can be further improved.

Author Contributions: Conceptualization, Z.H. and C.S.B.; investigation, Z.H.; writing—original draft preparation, Z.H.; writing—review and editing, L.S. (Letitia Stipkovits), L.S. (Luca Serventi), H.Z. and C.S.B.

Funding: This research received no external funding.

Conflicts of Interest: The authors declare no conflicts of interest.

References

1. Liu, Q.; Guo, W.; Zhu, X. Effect of lactose content on dielectric properties of whole milk and skim milk. *Int. J. Food Sci. Technol.* **2018**, *53*, 2037–2044. [CrossRef]
2. USDA National Nutrient Database for Standard Reference 1 April 2018 Software v.3.9.5.3. Available online: https://ndb.nal.usda.gov/ndb/search/list (accessed on 9 August 2019).
3. Ali, A.H.; Wei, W.; Abed, S.M.; Korma, S.A.; Mousa, A.H.; Hassan, H.M.; Jin, Q.; Wang, X. Impact of technological processes on buffalo and bovine milk fat crystallization behavior and milk fat globule membrane phospholipids profile. *LWT Food Sci. Technol.* **2018**, *90*, 424–432. [CrossRef]
4. Mattice, K.D.; Marangoni, A.G. Matrix effects on the crystallization behaviour of butter and roll-in shortening in laminated bakery products. *Food Res. Int.* **2017**, *96*, 54–63. [CrossRef] [PubMed]
5. Liu, H.; Hebb, R.L.; Putri, N.; Rizvi, S.S. Physical properties of supercritical fluid extrusion products composed of milk protein concentrate with carbohydrates. *Int. J. Food Sci. Technol.* **2018**, *53*, 847–856. [CrossRef]

6. Pareyt, B.; Finnie, S.M.; Putseys, J.A.; Delcour, J.A. Lipids in bread making: Sources, interactions, and impact on bread quality. *J. Cereal Sci.* **2011**, *54*, 266–279. [CrossRef]
7. Colla, K.; Costanzo, A.; Gamlath, S. Fat replacers in baked food products. *Foods* **2018**, *7*, 192. [CrossRef] [PubMed]
8. Rios, R.V.; Pessanha, M.D.F.; de Almeida, P.F.; Viana, C.L.; da Silva Lannes, S.C. Application of fats in some food products. *Food Sci. Technol.* **2014**, *34*, 3–15. [CrossRef]
9. Ghotra, B.S.; Dyal, S.D.; Narine, S.S. Lipid shortenings: A review. *Food Res. Int.* **2002**, *35*, 1015–1048. [CrossRef]
10. van Nieuwenhuyzen, W.; Beghin-Say, E. Lecithin specialities for baked goods. *Eur. Food Drink Rev.* **2001**, *1*, 37–41.
11. Tebben, L.; Shen, Y.; Li, Y. Improvers and functional ingredients in whole wheat bread: A review of their effects on dough properties and bread quality. *Trends Food Sci. Technol.* **2018**, *81*, 10–24. [CrossRef]
12. Kaylegian, K.E. Functional characteristics and nontraditional applications of milk Lipid components in food and nonfood systems. *J. Dairy Sci.* **1995**, *78*, 2524–2540. [CrossRef]
13. Stampfli, L.; Nersten, B. Emulsifiers in bread making. *Food Chem.* **1995**, *52*, 353–360. [CrossRef]
14. Winkler-Moser, J.K.; Mehta, B.M. Chemical composition of fat and oil products. In *Handbook of Food Chemistry*; Cheung, P.C.K., Mehta, B.M., Eds.; Springer: Berlin/Heidelberg, Germany, 2015; pp. 365–402.
15. Ortega-Anaya, J.; Jiménez-Flores, R. Symposium review: The relevance of bovine milk phospholipids in human nutrition–Evidence of the effect on infant gut and brain development. *J. Dairy Sci.* **2018**, *102*, 1–11. [CrossRef] [PubMed]
16. Arranz, E.; Corredig, M. Invited review: Milk phospholipid vesicles, their colloidal properties, and potential as delivery vehicles for bioactive molecules. *J. Dairy Sci.* **2017**, *100*, 4213–4222. [CrossRef] [PubMed]
17. Zhao, L.; Du, M.; Gao, J.; Zhan, B.; Mao, X. Label-free quantitative proteomic analysis of milk fat globule membrane proteins of yak and cow and identification of proteins associated with glucose and lipid metabolism. *Food Chem.* **2019**, *275*, 59–68. [CrossRef] [PubMed]
18. Nakatani, M.; Inoue, R.; Tomonaga, S.; Fukuta, K.; Tsukahara, T. Production, absorption, and blood flow dynamics of short-chain fatty acids produced by fermentation in piglet hindgut during the suckling-weaning period. *Nutrients* **2018**, *10*, 1220. [CrossRef] [PubMed]
19. Jung, H.J.; Ho, M.J.; Ahn, S.; Han, Y.T.; Kang, M.J. Synthesis and physicochemical evaluation of entecavir-fatty acid conjugates in reducing food effect on intestinal absorption. *Molecules* **2018**, *23*, 731. [CrossRef]
20. Bourlieua, C.; Cheillan, D.; Blota, M.; Daira, P.; Trauchessec, M.; Ruet, S.; Gassi, J.-Y.; Beaucher, E.; Robert, B.; Leconte, N.; et al. Polar lipid composition of bioactive dairy co-products buttermilk and butterserum: Emphasis on sphingolipid and ceramide isoforms. *Food Chem.* **2018**, *240*, 67–74. [CrossRef] [PubMed]
21. Cheema, M.; Smith, P.B.; Patterson, A.D.; Hristov, A.; Hart, F.M. The association of lipophilic phospholipids with native bovine casein micelles in skim milk: Effect of lactation stage and casein micelle size. *J. Dairy Sci.* **2017**, *101*, 8672–8687. [CrossRef] [PubMed]
22. Ferreiro, T.; Martínez, S.; Gayoso, L.; Rodríguez-Otero, J.L. Evolution of phospholipid contents during the production of quark cheese from buttermilk. *J. Dairy Sci.* **2016**, *99*, 4154–4159. [CrossRef] [PubMed]
23. Claumarchirant, L.; Cilla, A.; Matencio, E.; Sanchez-Siles, L.M.; Castro-Gomez, P.; Fontecha, J.; Alegría, A.; Lagarda, M.J. Addition of milk fat globule membrane as an ingredient of infant formulas for resembling the polar lipids of human milk. *Int. Dairy J.* **2016**, *61*, 228–238. [CrossRef]
24. Rodríguez-Alcal, L.M.; Castro-Gomez, P.; Felipe, X.; Noriega, L.; Fontecha, J. Effect of processing of cowmilk by high pressures under conditions up to 900 MPa on the composition of neutral, polar lipids and fatty acids. *LWT Food Sci. Technol.* **2015**, *62*, 265–270. [CrossRef]
25. Timms, R.E. Chapter 1 Physical chemistry of fats. In *Fats in Food Products*; Moran, D.P.J., Rajah, K.K., Eds.; Chapman & Hall, Springer Science+Business Media Dordrecht: Northampton, UK, 1994; pp. 1–28.
26. Latip, R.A.; Lee, Y.Y.; Tang, T.K.; Phuah, E.T.; Tan, C.P.; Lai, O.M. Physicochemical properties and crystallisation behaviour of bakery shortening produced from stearin fraction of palm-based diacyglycerol blended with various vegetable oils. *Food Chem.* **2013**, *141*, 3938–3946. [CrossRef] [PubMed]
27. Sciarini, L.S.; Van Bockstaele, F.; Nusantoro, B.; Pérez, G.T.; Dewettinck, K. Properties of sugar-snap cookies as influenced by lauric-based shortenings. *J. Cereal Sci.* **2013**, *58*, 234–240. [CrossRef]
28. Zhu, T.; Zhang, X.; Wu, H.; Li, B. Comparative study on crystallization behaviors of physical blend- and interesterified blend-based special fats. *J. Food Eng.* **2019**, *241*, 33–40. [CrossRef]

29. Wassell, P. Bakery fats. In *Fats in Food Technology*, 2nd ed.; Rajah, K.K., Ed.; John Wiley & Sons, Ltd.: Oxford, UK, 2014; pp. 39–81.
30. Meng, Z.; Guo, Y.; Wang, Y.; Liu, Y. Oleogels from sodium stearoyl lactylate-based lamellar crystals: Structural characterization and bread application. *Food Chem.* **2019**, *292*, 134–142. [CrossRef]
31. Fu, Y.; Zhao, R.; Zhang, L.; Bi, Y.; Zhang, H.; Chen, C. Influence of acylglycerol emulsifier structure and composition on the function of shortening in layer cake. *Food Chem.* **2018**, *249*, 213–221. [CrossRef]
32. Lee, C.L.; Liao, H.L.; Lee, W.C.; Hsu, C.K.; Hsueh, F.C.; Pan, J.Q.; Chu, C.H.; Wei, C.T.; Chen, M.J. Standards and labeling of milk fat and spread products in different countries. *J. Food Drug Anal.* **2018**, *26*, 469–480. [CrossRef]
33. Lopez, C.; Blot, M.; Briard-Bion, V.; Cirie, C.; Graulet, B. Butter serums and buttermilks as sources of bioactive lipids from the milk fat globule membrane: Differences in their lipid composition and potentialities of cow diet to increase n-3 PUFA. *Food Res. Int.* **2017**, *100*, 864–872. [CrossRef]
34. Yan, M.; Holden, N.M. Life cycle assessment of multi-product dairy processing using Irish butter and milk powders as an example. *J. Clean. Prod.* **2018**, *198*, 215–230. [CrossRef]
35. Chandan, R.C. Chapter 18 Dairy ingredients in bakery, snacks, sauces, dressings, processed meats, and functional Foods. In *Dairy Ingredients for Food Processing*; Chandan, R.C., Kilara, A., Eds.; Blackwell Publishing Ltd., John Wiley & Sons, Inc.: Chicago, IL, USA, 2011; pp. 473–500.
36. Antony, B.; Sharma, S.; Mehta, B.M.; Ratnam, K.; Aparnathi, K.D. Study of Fourier transform near infrared (FT-NIR) spectra of ghee (anhydrous milk fat). *Int. J. Dairy Technol.* **2018**, *71*, 484–490. [CrossRef]
37. Li, Y.; Zhao, J.; Xie, X.; Zhang, Z.; Zhang, N.; Wang, Y. A low trans margarine fat analog to beef tallow for healthier formulations: Optimization of enzymatic interesterification using soybean oil and fully hydrogenated palm oil. *Food Chem.* **2018**, *255*, 405–413. [CrossRef] [PubMed]
38. Lakum, R.; Sonwai, S. Production of trans-free margarine fat by enzymatic interesterification of soy bean oil, palm stearin and coconut stearin blend. *Int. J. Food Sci. Technol.* **2018**, *53*, 2761–2769. [CrossRef]
39. Chandan, R.C. Chapter 1 Dairy Ingredients for Food Processing: An Overview. In *Dairy Ingredients for Food Processing*; Chandan, R.C., Kilara, A., Eds.; Blackwell Publishing Ltd., John Wiley & Sons, Inc.: Chicago, IL, USA, 2011; pp. 3–33.
40. Ningtyas, D.W.; Bhandari, B.; Bansal, N.; Prakash, S. Effect of homogenisation of cheese milk and high-shear mixing of the curd during cream cheese manufacture. *Int. J. Dairy Technol.* **2018**, *71*, 417–431. [CrossRef]
41. Friedberg, J. Demanding more from dairy. *Snack Food Wholes. Bak.* **2018**, *107*, 26–29.
42. Ali, B.; Khan, K.Y.; Majeed, H.; Xu, L.; Bakry, A.M.; Raza, H.; Shoaib, M.; Wu, F.; Xu, X. Production of ingredient type flavoured white enzyme modified cheese. *J. Food Sci. Technol.* **2019**, *56*, 1683–1695. [CrossRef]
43. Karaca, Y. Production and quality of kefir cultured butter. *Mljekarstvo* **2018**, *68*, 64–72. [CrossRef]
44. Gemelas, L.; Degraeve, P.; Hallier, A.; Demarigny, Y. Fermented dairy product for a low-fat bakery product application: Chemical and sensory analysis. *Czech J. Food Sci.* **2016**, *34*, 529–533. [CrossRef]
45. Yu, J.; Mo, L.; Pan, L.; Yao, C.; Ren, D.; An, X.; Tsogtgerel, T.; Zhang, H.; Liu, W. Bacterial microbiota and metabolic character of traditional sour cream and butter in Buryatia, Russia. *Front. Microbiol.* **2018**, *9*, 2496. [CrossRef]
46. Wansink, B.; Linder, L.R. Interactions between forms of fat consumption and restaurant bread consumption. *Int. J. Obes. Relat. Metab. Disord.* **2003**, *27*, 866–868. [CrossRef]
47. Lai, H.-M.; Lin, T.C. Bakery products: Science and technology. In *Bakery Products: Science and Technology*, 2nd ed.; Hui, Y.H., Ed.; Blackwell Publishing: Chicago, IL, USA, 2014; Volume 43, pp. 3–68.
48. Xu, Y.; Zhu, X.; Ma, X.; Xiong, H.; Zeng, Z.; Peng, H.; Hu, J. Enzymatic production of trans-free shortening from coix seed oil, fully hydrogenated palm oil and Cinnamomum camphora seed oil. *Food Biosci.* **2018**, *22*, 1–8. [CrossRef]
49. Jukić, M.; Lukinac, J.; Čuljak, J.; Pavlović, M.; Šubarić, D.; Koceva Komlenić, D. Quality evaluation of biscuits produced from composite blends of pumpkin seed oil press cake and wheat flour. *Int. J. Food Sci. Technol.* **2019**, *54*, 602–609. [CrossRef]
50. Mert, B.; Demirkesen, I. Reducing saturated fat with oleogel/shortening blends in a baked product. *Food Chem.* **2016**, *199*, 809–816. [CrossRef] [PubMed]
51. Dalgleish, D.G. Chapter 1 Food emulsions: Their structures and properties. In *Food Emulsions*, 4th ed.; Friberg, S., Larsson, K., Sjoblom, J., Eds.; Marcel Dekker, Inc.: New York, NY, USA, 2004; pp. 1–60.

52. de la Horra, A.E.; Barrera, G.N.; Steffolani, E.M.; Ribotta, P.D.; Leon, A.E. Relationships between structural fat properties with sensory, physical and textural attributes of yeast-leavened laminated salty baked product. *J. Food Sci. Technol.* **2017**, *54*, 2613–2625. [CrossRef] [PubMed]
53. Brooker, B.E. The stabilisation of air in foods containing fat—A review. *Food Struct.* **1993**, *12*, 1–9.
54. Chin, N.L.; Rahman, R.A.; Hashim, D.M.; Kowng, S.Y. Palm oil shortening effects on baking performance of white bread. *J. Food Process. Eng.* **2010**, *33*, 413–433. [CrossRef]
55. Brooker, B.E. The role of fat in the stabilisation of gas cells in bread dough. *J. Cereal Sci.* **1996**, *24*, 187–198. [CrossRef]
56. Desai, S.P.I.; Naladala, S.; Anandharamakrishnan, C. Impact of wheat bran addition on the temperature-induced state transitions in dough during bread-baking process. *Int. J. Food Sci. Technol.* **2018**, *53*, 404–411.
57. Watanabe, A.; Yokomizo, K.; Eliasson, A.C. Effect of physical states of nonpolar lipids on rheology, ultracentrifugation, and microstructure of wheat flour dough. *Cereal Chem.* **2003**, *80*, 281–284. [CrossRef]
58. Devi, A.; Khatkar, B.S. Effects of fatty acids composition and microstructure properties of fats and oils on textural properties of dough and cookie quality. *J. Food Sci. Technol.* **2018**, *55*, 321–330. [CrossRef] [PubMed]
59. Gray, J.A.; Bemiller, J.N. Bread staling: Molecular basis and control. *Compr. Rev. Food Sci. Food Saf.* **2003**, *2*, 1–21. [CrossRef]
60. Huang, L.; Huang, Z.; Zhang, Y.; Zhou, S.; Hu, W.; Dong, M. Impact of tempeh flour on the rheology of wheat flour dough and bread staling. *LWT Food Sci. Technol.* **2019**, *111*, 694–702. [CrossRef]
61. Hesso, N.; Le-Bail, A.; Loisel, C.; Chevallier, S.; Pontoire, B.; Queveau, D.; Le-Bail, P. Monitoring the crystallization of starch and lipid components of the cake crumb during staling. *Carbohydr. Polym.* **2015**, *133*, 533–538. [CrossRef] [PubMed]
62. Mattice, K.D.; Marangoni, A.G. Gelatinized wheat starch influences crystallization behaviour and structure of roll-in shortenings in laminated bakery products. *Food Chem.* **2018**, *243*, 396–402. [CrossRef] [PubMed]
63. Chao, C.; Yu, J.; Wang, S.; Copeland, L.; Wang, S. Mechanisms underlying the formation of complexes between maize starch and lipids. *J. Agric. Food Chem.* **2018**, *66*, 272–278. [CrossRef] [PubMed]
64. Zhao, Y.; Khalid, N.; Shu, G.; Neves, M.A.; Kobayashi, I.; Nakajima, M. Complex coacervates from gelatin and octenyl succinic anhydride modified kudzu starch: Insights of formulation and characterization. *Food Hydrocoll.* **2019**, *86*, 70–77. [CrossRef]
65. Reddy, C.K.; Choi, S.M.; Lee, D.J.; Lim, S.T. Complex formation between starch and stearic acid: Effect of enzymatic debranching for starch. *Food Chem.* **2018**, *244*, 136–142. [CrossRef] [PubMed]
66. Li, X.; Gao, X.; Lu, J.; Mao, X.; Wang, Y.; Feng, D.; Cao, J.; Huang, L.; Gao, W. Complex formation, physicochemical properties of different concentration of palmitic acid yam (Dioscorea pposita Thunb.) starch preparation mixtures. *LWT Food Sci. Technol.* **2019**, *101*, 130–137. [CrossRef]
67. Gunenc, A.; Kong, L.; Elias, R.J.; Ziegler, G.R. Inclusion complex formation between high amylose corn starch and alkylresorcinols from rye bran. *Food Chem.* **2018**, *259*, 1–6. [CrossRef]
68. Mohebbi, Z.; Homayouni, A.; Azizi, M.H.; Hosseini, S.J. Effects of beta-glucan and resistant starch on wheat dough and prebiotic bread properties. *J. Food Sci. Technol.* **2018**, *55*, 101–110. [CrossRef]
69. Wei, C.; Fu, J.; Liu, D.; Zhang, Z.; Liu, G. Functional properties of chicken fat-based shortenings: Effects of based oils and emulsifiers. *Int. J. Food Prop.* **2018**, *20* (Suppl. 3), S3277–S3288. [CrossRef]
70. Garzon, R.; Hernando, I.; Llorca, E.; Rosell, C.M. Understanding the effect of emulsifiers on bread aeration during breadmaking. *J. Sci. Food Agric.* **2018**, *98*, 5494–5502. [CrossRef]
71. AACC. *Approved Methods of the AACC*, 11th ed.; American Association of Cereal Chemists (AACC): St. Paul, MN, USA, 2010.
72. Guo, X.; Sun, X.; Zhang, Y.; Wang, R.; Yan, X. Interactions between soy protein hydrolyzates and wheat proteins in noodle making dough. *Food Chem.* **2018**, *245*, 500–507. [CrossRef]
73. Fernandes, S.S.; Salas-Mellado, M.L. Addition of chia seed mucilage for reduction of fat content in bread and cakes. *Food Chem.* **2017**, *227*, 237–244. [CrossRef]
74. Graça, C.; Fradinho, P.; Sousa, I.; Raymundo, A. Impact of Chlorella vulgaris on the rheology of wheat flour dough and bread texture. *LWT Food Sci. Technol.* **2018**, *89*, 466–474. [CrossRef]
75. Smith, P.R.; Johansson, J. Influences of the proportion of solid fat in a shortening on loaf volume and staling of bread. *J. Food Process. Preserv.* **2004**, *28*, 359–367. [CrossRef]

76. O'brien, R.D. *Fats and Oils: Formulating and Processing for Applications*; CRC Press LLC: Boca Raton, FL, USA, 2004; p. 574.
77. Messia, M.C.; Reale, A.; Maiuro, L.; Candigliota, T.; Sorrentino, E.; Marconi, E. Effects of pre-fermented wheat bran on dough and bread characteristics. *J. Cereal Sci.* **2016**, *69*, 138–144. [CrossRef]
78. Sroan, B.S.; MacRitchie, F. Mechanism of gas cell stabilization in breadmaking. II. The secondary liquid lamellae. *J. Cereal Sci.* **2009**, *49*, 41–46. [CrossRef]
79. Gerits, L.R.; Pareyt, B.; Delcour, J.A. A lipase based approach for studying the role of wheat lipids in bread making. *Food Chem.* **2014**, *156*, 190–196. [CrossRef]
80. Chong, H.M.; Mohammed, I.K.; Linter, B.; Allen, R.; Charalambides, M.N. Mechanical and microstructural changes of cheese cracker dough during baking. *LWT Food Sci. Technol.* **2017**, *86*, 148–158. [CrossRef]
81. Purhagen, J.K.; Sjöö, M.E.; Eliasson, A.-C. Starch affecting anti-staling agents and their function in freestanding and pan-baked bread. *Food Hydrocoll.* **2011**, *25*, 1656–1666. [CrossRef]
82. Sowmya, M.; Jeyarani, T.; Jyotsna, R.; Indrani, D. Effect of replacement of fat with sesame oil and additives on rheological, microstructural, quality characteristics and fatty acid profile of cakes. *Food Hydrocoll.* **2009**, *23*, 1827–1836. [CrossRef]
83. Sahin, A.W.; Rice, T.; Zannini, E.; Lynch, K.M.; Coffey, A.; Arendt, E.K. The incorporation of sourdough in sugar-reduced biscuits: A promising strategy to improve techno-functional and sensory properties. *Eur. Food Res. Technol.* **2019**, *2019*, 1–14. [CrossRef]
84. Han, J.; Janz, J.A.M.; Gerlat, M. Development of gluten-free cracker snacks using pulse flours and fractions. *Food Res. Int.* **2010**, *43*, 627–633. [CrossRef]
85. Blanco Canalis, M.S.; Valentinuzzi, M.C.; Acosta, R.H.; León, A.E.; Ribotta, P.D. Effects of Fat and Sugar on Dough and Biscuit Behaviours and their Relationship to Proton Mobility Characterized by TD-NMR. *Food Bioprocess. Technol.* **2018**, *11*, 953–965. [CrossRef]
86. Martinez-Saez, N.; Hochkogler, C.M.; Somoza, V.; Del Castillo, M.D. Biscuits with No Added Sugar Containing Stevia, Coffee Fibre and Fructooligosaccharides Modifies alpha-Glucosidase Activity and the Release of GLP-1 from HuTu-80 Cells and Serotonin from Caco-2 Cells after In Vitro Digestion. *Nutrients* **2017**, *9*, 694. [CrossRef]
87. Sadowska-Rociek, A.; Cieslik, E. Carbohydrate-based fat mimetics can affect the levels of 3-monochloropropane-1,2-diol esters and glycidyl esters in shortbread biscuits. *Plant Foods Hum. Nutr.* **2019**, *74*, 216–222. [CrossRef]
88. Ronda, F.; Oliete, B.; Gómez, M.; Caballero, P.A.; Pando, V. Rheological study of layer cake batters made with soybean protein isolate and different starch sources. *J. Food Eng.* **2011**, *102*, 272–277. [CrossRef]
89. UMI. *Reference Manual for U.S. Milk Powders and Microfiltered Ingredients: Think USA Dairy*; USA Dairy Export Council (USDEC); Dairy Management Inc. (DMI): Arlington, VA, USA, 2018; p. 135.
90. Palav, T.S. Chemistry of cake manufacturing. In *Reference Module in Food Sciences*; Smithers, G.W., Ed.; Elsevier Inc.: Amsterdam, The Netherlands, 2016; pp. 1–8.
91. Zhou, J.; Faubion, J.M.; Walker, C.E. Evaluation of different types of fats for use in high-ratio layer cakes. *LWT Food Sci. Technol.* **2011**, *44*, 1802–1808. [CrossRef]
92. Nhouchi, Z.; Botosoa, E.P.; Chene, C.; Karoui, R. Potentiality of front-face fluorescence and mid-infrared spectroscopies coupled with partial least square regression to predict lipid oxidation in pound cakes during storage. *Food Chem.* **2019**, *275*, 322–332. [CrossRef]
93. Bajaj, R.; Singh, N.; Kaur, A. Effect of native and gelatinized starches from various sources on sponge cake making characteristics of wheat flour. *J. Food Sci. Technol.* **2019**, *56*, 1046–1055. [CrossRef]
94. Kozlowska, M.; Zbikowska, A.; Szpicer, A.; Poltorak, A. Oxidative stability of lipid fractions of sponge-fat cakes after green tea extracts application. *J. Food Sci. Technol.* **2019**, *56*, 2628–2638. [CrossRef]
95. BNF. *What is Energy Density?* British Nutrition Foundation: London, UK, 2018.
96. Serinyel, G.; Öztürk, S. Investigation on potential utilization of native and modified starches containing resistant starch as a fat replacer in bakery products. *Starch Stärke* **2017**, *69*, 1–9. [CrossRef]
97. Biguzzi, C.; Schlich, P.; Lange, C. The impact of sugar and fat reduction on perception and liking of biscuits. *Food Qual. Prefer.* **2014**, *35*, 41–47. [CrossRef]
98. Oh, I.K.; Amoah, C.; Lim, J.; Jeong, S.; Lee, S. Assessing the effectiveness of wax-based sunflower oil oleogels in cakes as a shortening replacer. *LWT Food Sci. Technol.* **2017**, *86*, 430–437. [CrossRef]

99. Kim, J.Y.; Lim, J.; Lee, J.; Hwang, H.S.; Lee, S. Utilization of oleogels as a replacement for solid fat in aerated baked goods: Physicochemical, rheological, and tomographic characterization. *J. Food Sci.* **2017**, *82*, 445–452. [CrossRef]
100. Jang, A.; Bae, W.; Hwang, H.S.; Lee, H.G.; Lee, S. Evaluation of canola oil oleogels with candelilla wax as an alternative to shortening in baked goods. *Food Chem.* **2015**, *187*, 525–529. [CrossRef]
101. Rodríguez-García, J.; Puig, A.; Salvador, A.; Hernando, I. Optimization of a sponge cake formulation with inulin as fat replacer: Structure, physicochemical, and sensory properties. *J. Food Sci.* **2012**, *77*, C189–C197. [CrossRef]
102. Błońska, A.; Marzec, A.; Błaszczyk, A. Instrumental evaluation of acoustic and mechanical texture properties of short-dough biscuits with different content of fat and inulin. *J. Texture Stud.* **2014**, *45*, 226–234. [CrossRef]
103. Rodríguez-García, J.; Laguna, L.; Puig, A.; Salvador, A.; Hernando, I. Effect of fat replacement by inulin on textural and structural properties of short dough biscuits. *Food Bioprocess. Technol.* **2013**, *6*, 2739–2750. [CrossRef]
104. Lee, Y.; Puligundla, P. Characteristics of reduced-fat muffins and cookies with native and modified rice starches. *Emir. J. Food Agric.* **2016**, *28*, 311. [CrossRef]
105. Krystyjan, M.; Gumul, D.; Ziobro, R.; Sikora, M. The effect of inulin as a fat replacement on dough and biscuit properties. *J. Food Qual.* **2015**, *38*, 305–315. [CrossRef]
106. Forker, A.; Zahn, S.; Rohm, H. A combination of fat replacers enables the production of fat-reduced shortdough biscuits with high-sensory quality. *Food Bioprocess. Technol.* **2011**, *5*, 2497–2505. [CrossRef]
107. Pehlivanoglu, H.; Ozulku, G.; Yildirim, R.M.; Demirci, M.; Toker, O.S.; Sagdic, O. Investigating the usage of unsaturated fatty acid-rich and low-calorie oleogels as a shortening mimetics in cake. *J. Food Process. Preserv.* **2018**, *42*, 1–11. [CrossRef]
108. Onacik-Gür, S.; Żbikowska, A.; Jaroszewska, A. Effect of high-oleic sunflower oil and other pro-health ingredients on physical and sensory properties of biscuits. *CyTA J. Food* **2015**, *13*, 621–628. [CrossRef]
109. Felisberto, M.H.F.; Wahanik, A.L.; Gomes-Ruffi, C.R.; Clerici, M.T.P.S.; Chang, Y.K.; Steel, C.J. Use of chia (Salvia hispanica L.) mucilage gel to reduce fat in pound cakes. *LWT Food Sci. Technol.* **2015**, *63*, 1049–1055. [CrossRef]
110. Romanchik-Cerpovicz, J.E.; Jeffords, M.J.A.; Onyenwoke, A.C. College student acceptance of chocolate bar cookies containing puree of canned green peas as a fat-ingredient substitute. *J. Culin. Sci. Technol.* **2018**, *2018*, 1–12. [CrossRef]
111. Domenech-Asensi, G.; Merola, N.; Lopez-Fernandez, A.; Ros-Berruezo, G.; Frontela-Saseta, C. Influence of the reformulation of ingredients in bakery products on healthy characteristics and acceptability of consumers. *Int. J. Food Sci. Nutr.* **2016**, *67*, 74–82. [CrossRef]
112. Giarnetti, M.; Paradiso, V.M.; Caponio, F.; Summo, C.; Pasqualone, A. Fat replacement in shortbread cookies using an emulsion filled gel based on inulin and extra virgin olive oil. *LWT Food Sci. Technol.* **2015**, *63*, 339–345. [CrossRef]
113. Laneuville, S.I.; Paquin, P.; Turgeon, S.L. Formula optimization of a low-fat food system containing whey protein isolate- xanthan gum complexes as fat replacer. *J. Food Sci.* **2005**, *70*, s513–s519. [CrossRef]
114. Gerits, L.R.; Pareyt, B.; Masure, H.G.; Delcour, J.A. A lipase based approach to understand the role of wheat endogenous lipids in bread crumb firmness evolution during storage. *LWT Food Sci. Technol.* **2015**, *64*, 874–880. [CrossRef]
115. Gerits, L.R.; Pareyt, B.; Masure, H.G.; Delcour, J.A. Native and enzymatically modified wheat (Triticum aestivum L.) endogenous lipids in bread making: A focus on gas cell stabilization mechanisms. *Food Chem.* **2015**, *172*, 613–621. [CrossRef]
116. Scheuer, P.M.; Mattioni, B.; Barreto, P.L.M.; Montenegro, F.M.; Gomes-Ruffi, C.R.; Biondi, S.; Kilpp, M.; de Francisco, A. Effects of fat replacement on properties of whole wheat bread. *Braz. J. Pharm. Sci.* **2014**, *50*, 703–712. [CrossRef]
117. Sharp, T. Technical constraints in the development of reduced-fat bakery products. *Proc. Nutr. Soc.* **2001**, *60*, 489–496. [CrossRef]
118. Dapčević Hadnađev, T.; Hadnađev, M.; Pojić, M.; Rakita, S.; Krstonošić, V. Functionality of OSA starch stabilized emulsions as fat replacers in cookies. *J. Food Eng.* **2015**, *167*, 133–138. [CrossRef]

119. Serin, S.; Sayar, S. The effect of the replacement of fat with carbohydrate-based fat replacers on the dough properties and quality of the baked pogaca: A traditional high-fat bakery product. *Food Sci. Technol.* **2016**, *37*, 25–32. [CrossRef]
120. Lim, J.; Ko, S.; Lee, S. Use of Yuja (Citrus junos) pectin as a fat replacer in baked foods. *Food Sci. Biotechnol.* **2014**, *23*, 1837–1841. [CrossRef]
121. Santiago-García, P.A.; Mellado-Mojica, E.; León-Martínez, F.M.; López, M.G. Evaluation of Agave angustifolia fructans as fat replacer in the cookies manufacture. *LWT Food Sci. Technol.* **2017**, *77*, 100–109. [CrossRef]
122. Lindarte Artunduaga, J.; Gutierrez, L.F. Effects of replacing fat by betaglucans from Ganoderma lucidum on batter and cake properties. *J. Food Sci. Technol.* **2019**, *56*, 451–461. [CrossRef]
123. Seppänen-Laakso, T.; Vanhanen, H.; Laakso, I.; Kohtamäki, H.; Viikari, J. Replacement of butter on bread by rapeseed oil and rapeseed oil-containing margarine: Effects on plasma fatty acid composition and serum cholesterol. *Br. J. Nutr.* **2007**, *68*, 639–654. [CrossRef]
124. Liu, Y.; Meng, Z.; Shan, L.; Jin, Q.; Wang, X. Preparation of specialty fats from beef tallow and canola oil by chemical interesterification: Physico-chemical properties and bread applications of the products. *Eur. Food Res. Technol.* **2009**, *230*, 457–466. [CrossRef]
125. Jacob, J.; Leelavathi, K. Effect of fat-type on cookie dough and cookie quality. *J. Food Eng.* **2007**, *79*, 299–305. [CrossRef]
126. Zhang, Z.; Shim, Y.Y.; Ma, X.; Huang, H.; Wang, Y. Solid fat content and bakery characteristics of interesterified beef tallow-palm mid fraction based margarines. *RSC Adv.* **2018**, *8*, 12390–12399. [CrossRef]
127. Basso, R.C.; Ribeiro, A.P.B.; Masuchi, M.H.; Gioielli, L.A.; Gonçalves, L.A.G.; dos Santos, A.O.; Cardoso, L.P.; Grimaldi, R. Tripalmitin and monoacylglycerols as modifiers in the crystallisation of palm oil. *Food Chem.* **2010**, *122*, 1185–1192. [CrossRef]
128. Calligaris, S.; Manzocco, L.; Valoppi, F.; Nicoli, M.C. Effect of palm oil replacement with monoglyceride organogel and hydrogel on sweet bread properties. *Food Res. Int.* **2013**, *51*, 596–602. [CrossRef]
129. Manzocco, L.; Calligaris, S.; Da Pieve, S.; Marzona, S.; Nicoli, M.C. Effect of monoglyceride-oil–water gels on white bread properties. *Food Res. Int.* **2012**, *49*, 778–782. [CrossRef]
130. Singh, A.; Auzanneau, F.I.; Rogers, M.A. Advances in edible oleogel technologies—A decade in review. *Food Res. Int.* **2017**, *97*, 307–317. [CrossRef]
131. Martins, A.J.; Cerqueira, M.A.; Fasolin, L.H.; Cunha, R.L.; Vicente, A.A. Beeswax organogels: Influence of gelator concentration and oil type in the gelation process. *Food Res. Int.* **2016**, *84*, 170–179. [CrossRef]
132. Tarancón, P.; Fiszman, S.M.; Salvador, A.; Tárrega, A. Formulating biscuits with healthier fats. Consumer profiling of textural and flavour sensations during consumption. *Food Res. Int.* **2013**, *53*, 134–140. [CrossRef]
133. Laguna, L.; Primo-Martín, C.; Varela, P.; Salvador, A.; Sanz, T. HPMC and inulin as fat replacers in biscuits: Sensory and instrumental evaluation. *LWT Food Sci. Technol.* **2014**, *56*, 494–501. [CrossRef]
134. Tarancón, P.; Salvador, A.; Sanz, T.; Fiszman, S.; Tárrega, A. Use of healthier fats in biscuits (olive and sunflower oil): Changing sensory features and their relation with consumers' liking. *Food Res. Int.* **2015**, *69*, 91–96. [CrossRef]

135. Hu, S.-M.; Lai, H.-S. Developing low-fat banana bread by Using okra gum as a fat replacer. *J. Culin. Sci. Technol.* **2016**, *15*, 36–42. [CrossRef]
136. Othman, N.A.; Abdul Manaf, M.; Harith, S.; Wan Ishak, W.R. Influence of avocado puree as a fat replacer on nutritional, fatty acid, and organoleptic properties of low-fat muffins. *J. Am. Coll. Nutr.* **2018**, *37*, 583–588. [CrossRef] [PubMed]

© 2019 by the authors. Licensee MDPI, Basel, Switzerland. This article is an open access article distributed under the terms and conditions of the Creative Commons Attribution (CC BY) license (http://creativecommons.org/licenses/by/4.0/).

Review

Using Pulses in Baked Products: Lights, Shadows, and Potential Solutions

Andrea Bresciani and Alessandra Marti *

Department of Food, Environmental, and Nutritional Sciences, Università degli Studi di Milano, Via G. Celoria 2, 20133 Milan, Italy; andrea.bresciani@unimi.it
* Correspondence: alessandra.marti@unimi.it

Received: 17 July 2019; Accepted: 17 September 2019; Published: 2 October 2019

Abstract: Nowadays, consumers are more conscious of the environmental and nutritional benefits of foods. Pulses—thanks to both nutritional and health-promoting features, together with their low environmental impact—satisfy the demand for high-protein/high-fiber products. However, their consumption is still somewhat limited in Western countries, due to the presence of antinutrient compounds including phytic acid, trypsin inhibitors, and some undigested oligosaccharides, which are responsible for digestive discomfort. Another limitation of eating pulses regularly is their relatively long preparation time. One way to increase the consumption of pulses is to use them as an ingredient in food formulations, such as bread and other baked products. However, some sensory and technological issues limit the use of pulses on an industrial scale; consequently, they require special attention when combined with cereal-based products. Developing formulations and/or processes to improve pulse quality is necessary to enhance their incorporation into baked products. In this context, this study provides an overview of strengths and weaknesses of pulse-enriched baked products focusing on the various strategies—such as the choice of suitable ingredients or (bio)-technological approaches—that counteract the negative effects of including pulses in baked goods.

Keywords: pulses; bread; bio-technological processing; air classification; fermentation; germination

---

1. Introduction

Legumes or pulses are dry edible seeds of plants belonging to the Fabaceae (*Leguminoseae*) family, which include field peas, dry beans, lentils, chickpeas and faba beans. The contemporary definition of pulses excludes oilseed legumes and legumes consumed in immature form [1]. Egypt and India consume the largest quantity of pulses; in these regions, pulses play a key role in helping the population to consume suitable levels of several important nutrients, particularly proteins, while in developed countries protein intake is mainly due to the consumption of animal-derived proteins [2]. In Europe, 60% of pulses are consumed in Spain, France, and the UK. It is also important to consider that the way pulses are prepared varies depending on world regions [3]. Nonetheless, pulses are traditionally consumed whole or split after soaking and cooking, although recently they have become increasingly popular and are widely used in food products such as pasta, bread, and other bakery products [4]. Indeed, legumes or pulses represent one of the possible ways to help solve global food security challenges. Indeed, as an inexpensive, sustainable source of proteins and other key nutrients, pulses meet the nutrition and food security requirements of the global population and can support the creation of sustainable and stable agricultural production systems, which could limit the negative effects due to climate change. In this context, the year 2016 was designated by the United Nations as the International Year of Pulses. The purpose of this initiative was to increase public awareness of the nutritional benefits of pulses and their potential role in improving global food security. Three years later, the impact of the International Year of Pulses is still making itself felt.

Pulses are mainly consumed as a whole, but Western countries are increasingly using pulse flour in food preparations for the general population or followers of special diets such as vegetarian, vegan or gluten-free. However, the consumption of pulses is limited due to the presence of antinutrients such as phytic acid, trypsin inhibitors and some non-digestible oligosaccharides that are related, for example, to digestive discomfort [5]. Moreover, the presence of off-flavors discourages the consumption of pulses for some [6]. Therefore, if we want more people to enjoy the scientifically recognized nutritional and health benefits of pulses, it is necessary to find ways and means of improving their quality as ingredients in baked products. In this frame, this review presents:

- the agronomic, compositional, and nutritional benefits of pulses;
- the pro and cons of using pulses in baked products;
- the various approaches for counteracting the negative effects of including them in food formulations.

2. Agronomic, Compositional, and Nutritional Benefits

The awareness and demand for pulses is still growing, and new pulse-containing products are launched on the market every year to meet the demand for products that are gluten-free, high in proteins and fiber, with a low glycemic index and a clean label. Moreover, end-use applications of pulses have generated research interest in many disciplines, such as breeding, genetics, agronomics, health and nutrition. An overview of the scientific literature of the last ten years setting 'legumes' or 'pulses' as a search term, resulted in the identification of about 1934 scientific papers in the area of Food Science and Technology (Figure 1). The graph in Figure 1 highlights that the number of publications has constantly been increasing over the years (except for 2019, which is still in progress), suggesting that the interest in this topic is still growing. Moreover, the marked increase in publications since the 2016 should be noted, possibly because of Year of Pulses announcement by FAO.

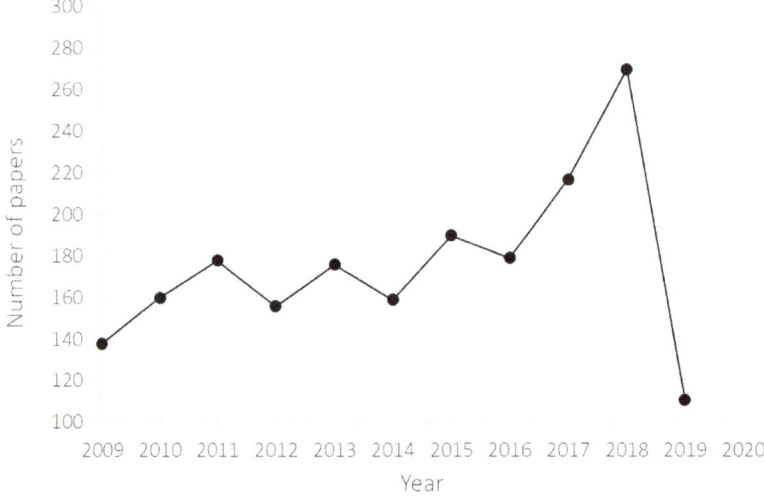

Figure 1. Papers on legumes in the field of food science and technology (source: Web of Science; 2009–2019; updated to 4 July 2019).

Taking into consideration all the pertinent research disciplines, more than 700 review papers have been published on pulses in the last five years. A tentative classification of the reviews published in the last two years according to their particular research area is summarized in Table 1; most concern plant science and agronomy (43%), while others (15%) are dedicated to the development of food products, including bread, pasta, snacks and cookies, enriched with pulses to improve their nutritional properties. Finally, the nutritional properties and the health benefits of pulses are the focus of about 10% of the total reviews.

Table 1. Topics of the main reviews published on pulses (source: Food Science and Technology Abstracts; 2018–2019; updated to 03 July 2019).

Research Area	Topic	Reference
Plant science/agronomy	Breeding	Pratap et al. [7]; Morris et al. [8]; Warsam et al. [9]
	Cultivation	Farooq et al. [10]
Nutrition/Health	Health benefits	Luna-Vita et al. [11]; Harouna et al. [12]; Harsha et al. [13]
	Bioactive compounds	Awika et al. [14]; Chhikara et al. [15]; Yi-Shen et al. [16]
	Allergens	Cabanillas et al. [17]
	Anti-nutritional factors	Avilés-Gaxiola et al. [18]
	Starch digestibility	Jeong et al. [19]
Processing	Milling	Thakur et al. [20]; Scanlon et al. [21]; Vishwakarma et al. [22]
	Enhancing nutritional properties	Van-der-Poe et al. [23]; Nkhata et al. [24]
	Bread fortification	Boukid et al. [25]; Rehman et al. [26]; Zhong et al. [27]
Functionality	General	Foschia et al. [28]; Jarpa-Parra [29]
	Emulsifiers	Burger et al. [30]; Sharif et al. [31]
	Structure-function relationship	Shevkani et al. [32]; Lam et al. [33]

2.1. Agronomic Traits

From an agronomic standpoint, pulses contribute nitrogen to the soil rather than extracting it from the soil. Many of them (e.g., lentils, field peas, and chickpeas) are also more drought- and temperature-tolerant than corn or soybeans. In addition, they tend to be disease-resistant and grow well in areas where weed pressure is low, thereby minimizing the need for pesticides and herbicides [34]. Pulses play an important role in improving soil fertility and would be of great importance for farmers with no or limited access to nitrogen fertilizers [35]. Cultivation of pulses and their use in food formulations instead of cereals could be an ideal approach to reduce and control the effects of climate change. In fact, pulse cultivation has low environmental impact, thanks to their low carbon and water footprints. Carbon footprints are mainly associated with agricultural greenhouse gas emissions [36]. The water footprint of pulses has much less impact compared to that of cereals and other protein sources, such as milk, chicken eggs and meat [37]. Their higher sustainability compared to cereals makes pulses appealing ingredients for food production.

2.2. Compositional Traits

Pulses have a different chemical composition compared to cereals: they are lower in carbohydrates (60–65%) but richer in proteins (21–25%) and fibers (12–20%), and for this reason, pulses are a suitable ingredient for the reformulation and enrichment of bread [5].

The primary storage carbohydrate of pulses is starch, which constitutes a major fraction of total carbohydrates for almost all the species [38]. Starch is composed of amylose and amylopectin. The ratio between these components depends on the starch structure. Generally, in cereals the quantity of amylose is about 18% even if values up to 30% have been observed in high amylose varieties (mainly rice and corn) [39]. Pulse starch provides health benefits because its high amylose content promotes the formation of resistant starch that cannot be hydrolyzed during digestion. Also, dietary fiber remains undigested in the small intestine, whereas it is fermented by the microbiota in the colon.

Colonic fermentation leads to the growth of beneficial bacteria and an increase in the production of short chain fatty acids, which have been associated with reduced risk of colon cancer [40]. High amylose content makes the starch structure more compact and limits its gelatinization capacity during heating. Moreover, starch remains partially crystalline during cooking due to its cell walls which remain undisrupted during heating. The preservation of the cell walls in whole pulses after cooking seems to prevent starch hydrolysis by digestive enzymes. This may contribute to both the high levels of resistant starch and low glycemic index of pulses [41].

As regards non-starch polysaccharides, pulses are a significant source of dietary fiber. There is a wide degree of variation in the amounts of dietary fiber as well as the ratio of soluble to insoluble fiber in pulses [42]. The total dietary fiber content in pulses range from 14 to 32% (dry weight) depending on the species [41]. There are various types of dietary fiber in pulses, including long chain soluble and insoluble polysaccharides, galacto-oligosaccharides and, as mentioned above, resistant starch. While insoluble fiber is generally combined with laxation, soluble fiber is linked with reducing cholesterol levels and ameliorating post-prandial blood glucose levels. Both soluble and insoluble fibers can act as prebiotics, supplying nutrients for gut microorganisms. Flours and fiber-rich fractions from pulses can be successfully utilized to increase the dietary fiber content (soluble and insoluble fiber) of processed foods, which has been shown to have health benefits. As for nutritional and health-promoting effects, pulse fibers can also be useful to improve the textural properties of foods by binding and retaining fat and/or moisture [42].

As regards sugars, monosaccharides generally make up less than 1% of pulse seed weight, whereas oligosaccharides make up 14%. In contrast to monosaccharides, oligosaccharides (i.e., raffinose and stachyose) are non-digestible by humans because of the β-glycosidic bond that links monosaccharides together and these oligosaccharides pass undigested through the stomach and upper intestine. In the lower intestine, they are fermented by gas-producing bacteria and make carbon dioxide, methane, and/or hydrogen, leading to the flatulence commonly associated with eating pulses [5].

Pulses contain relatively high amounts of proteins—about twice as much proteins as cereal grains—and for this reason, in many regions of the world, pulses are the major source of dietary protein and often represent a supplement for other protein sources. The most abundant storage proteins in pulses are globulins and albumins, which are classified as soluble proteins. Globulins (soluble in salt-water solutions) represent approximately 70% of the total proteins in pulses, whereas albumins (soluble in water) account for 10–20% [43]. Pulse proteins have low levels of sulfur amino acids, but the amount of lysine is greater than cereals. Therefore, pulse and cereal proteins are nutritionally complementary [44]. In addition to these nutritional features, pulse proteins are interesting from a technological standpoint due to their functional properties, including solubility, water holding capacity, and emulsifying and foaming properties that have been extensively described by several authors [30–33] as reported in Table 1.

Pulses are also rich sources of many micronutrients, including selenium, thiamin, niacin, folate, riboflavin, pyridoxine, potassium, zinc, vitamin E, and vitamin A [5]. In addition to being a good source of fiber, proteins, and micronutrients pulses are a source of compounds with antioxidant activity—i.e., flavonoid compounds, such as anthocyanins, quercetin glycosides, and proanthocyanidins, and isoflavones—that could limit the risk of certain diseases and promote overall health [45].

2.3. Potential Health Benefits

Consumption of pulses is encouraged in the diets of the general population, as their nutrient profile can have a positive health impact. Evidence of the link between the nutritional composition of pulses and the reduction in the risk of cardiovascular disease (CVD) and diabetes comes from more than 2000 studies published in the last ten years (source: Web of Science; updated to 09 September 2019). The potential health benefits of the consumption of beans [46], chickpeas [47], lentils [48], and peas [49] are summarized in Table 2. The intent of the authors was to provide a summary of the health benefits of pulses, not an extensive review. For more details about the nutritional and health benefits of pulses, we refer to the reading of recent reviews [5,50–52]. In collecting the information from the above-mentioned papers, the authors have noted that the majority of the studies focused on grains, while the health benefits of pulses-enriched baked goods have not been systematically evaluated, suggesting further studies to fill the current knowledge gap in this research area.

The high fiber and protein content of pulses has been shown to result in an increase in satiety and may contribute to decreasing the occurrence of obesity by reducing calorie intake and managing body weight over time [53]. These aspects have been shown to decrease the risk of developing type 2 diabetes and CVD [54].

Table 2. Potential health benefits of pulses and their mechanisms.

Health Benefits	Key Component	Mechanism	Reference
Colon cancer	Fiber	Anti-proliferative activity and inducing apoptosis in colon cancer cells	Mathers et al. [58]
Heart disease	Fiber	Reduction of blood pressure	Jayalath et al. [57]
	Mono- and polyunsaturated fat; sterols	Increase in high-density lipoprotein (HDL) cholesterol and decrease in both low-density lipoprotein (LDL) and total cholesterol	Bazzano et al. [55]
Diabetes	Resistant starch	Improvement of glucose tolerance as well as insulin sensitivity	Jenkins et al. [56]
Weight Control	Fiber	Interference with caloric intake by increasing chewing time and satiety	McCrory et al. [53]

Also, lipid profiles—sterols and mono-and polyunsaturated fats—contribute to reduce overall risk of CVD and atherosclerosis, decreasing total serum triglycerides and cholesterol [55].

Pulses are an ideal food choice for individuals with diabetes thanks to their low GI and high fiber content. Consuming pulses as part of a low-GI diet improved glycemic control, may help lower the risk of diabetes-related complications [56]. Additionally, insulin sensitivity and glucose tolerance are also improved with the presence of resistant starch that is high in pulses, which also helps reduce risks associated with diabetes. Therefore, it has been shown that the consumption of pulses is beneficial to the management of type 2 diabetes, metabolic syndrome and obesity. Pulse consumption is also linked to a reduction in cardiovascular disease and risk of cancer, they also contribute to overall health and wellness [57]. Moreover, the absence of gluten in pulses allows celiac disease sufferers to digest them.

The agronomic, compositional and nutritional benefits of pulses are the driving force for the growing interest in producing pulse-derived foods that are healthy, convenient, and rich in protein and fiber. In this context, they are usually processed to obtain flour or fiber, starch, and protein concentrates or isolates [59]. These different ingredients could be used in food reformulation to improve the physico-chemical, nutritional and technological properties of baked products.

3. Using Pulses in Baked Products

The growing interest in gluten-free, vegan and vegetarian diets has resulted in an increase in pulse consumption. Flour from pulses is mixed with other grains (with or without gluten) to make bread, biscuits or cookies, and other baked products. The following section will summarize the strengths and weaknesses of pulse-enriched foods in the past ten years.

3.1. Bread

Bread, a traditional and economical product that is easy to prepare and consume, is one of the most popular foods worldwide and is generally prepared from common wheat. Thus, it is a source of calories and of complex carbohydrates, with a modest amount of essential amino acids such as lysine and threonine. Using refined white flour instead of wholemeal flour, however, reduces the nutritional density and fiber content of white bread [60]. Nowadays consumers are more health oriented and conscious of the environmental and nutritional benefits of food. In response to consumer demands, the food industry is formulating vegetable-based products that fully satisfy the health and cultural concerns of today's typical consumer. From this point of view, pulses are a potential ingredient to improve the quality of products that are already widely consumed. Pulse-enriched wheat flour represents a potential way to increase the nutritional properties of cereal-based foods; it is well known that the amino acid composition of pulses complements that of cereals [43]. They are also rich in bioactive compounds, including fiber [61]. In addition, pulses are characterized by reduced starch bioavailability and high resistant starch content. Most of the studies are focused on reformulating wheat bread, mainly with lentils [62,63], chickpeas [62,64], and peas [62,65]. Protein concentrate and protein isolate from peas, lentils and chickpeas have been successfully incorporated in baked products [62]. However, using concentrated protein leads only to an increase in total protein content, losing the potential health benefits associated with other components present in the flour, including phenolic compounds, fiber, and minerals.

The incorporation of high amount of pulses has been successfully obtained in biscuits, cake, and other chemically leavened products (see section below). On the contrary, it has been a challenge to make bread, because gluten plays a structuring role in bread. On one hand, pulse proteins are not able to form gluten networks, on the other, weak interactions between pulse and wheat proteins reduce the formation of viscoelastic dough and affect air incorporation and gas retention during leavening, resulting in bread with poor crumb structure and texture. Thus, the addition of chickpea or peas flour is limited to percentages below 10–15% [64]. Generally, pulses are incorporated in common wheat flours, as recently reviewed by Boukid et al. [25]. The differences in observations among studies would most likely be due to differences in types of pulses (lentils, chickpea, etc.) and whether the pulses integrated in the formulation constitute dehulled or hulled flour. Unfortunately, most of the studies did not report any details about the type of flour, i.e., whether they were used after dehulling, making the comparison of the outcomes of different studies difficult. The presence of the structural fiber found in dehulled material would influence dough formation and bread performance.

Generally speaking, chickpea replacement of less than 10% creates some difficulties in dough preparation, including increased dough stickiness and reduced dough extensibility [66]. When more than 10% of wheat flour is replaced with fiber from peas, lentils, and chickpeas a significant decrease in water absorption is observed, which could be attributed to the higher amount of fiber present [67]. Moreover, the incorporation of flour from dehulled lentils decreases the time required to form the dough and its stability during mixing, together with its resistance to extension, likely due to gluten dilution [68]. Mixtures of wheat and dehulled lentil flours with 20% inclusion have high protein content but low water absorption, resulting in loaves with extremely reduced volumes and dense crumb structures [68]. Thus, high ratio pulse blends are indicated for different baked products such as biscuits (as described in the following section) or extruded products such as noodles. Blends of up to 15% pulse flour generally result in good loaf volume, firmness, and crumb structure. With increasing pulse levels, loaf volume incrementally decreases, and the color of the crumb darkens due to the Maillard reaction [63,66,68], resulting in a decrease in taste and overall acceptability [65]. Specifically, the best sensorial results in terms of appearance, taste, and color are obtained with the addition of up to 10% pulse flour (specifically peas) for bread, whereas higher proportions lead to a worsening of the product's sensory profile [65]. Finally, the addition of pulses (i.e., chickpeas and peas) has been shown to increase crumb firmness [64,69], likely due to their high amylose content compared to cereals, as mentioned above. The effects of adding pulses on dough rheology and bread quality are reviewed in detail by Boukid et al. [25] and Mohammed et al. [66].

3.2. Other Products

As regards biscuits, reformulation by adding pulses is not as challenging as for bread, since the formation of a gas-retaining gluten network during leavening and baking is not required. Moreover, the increase in hardness associated with pulse enrichment is not such a concern as in bread and could even be positive in cookies.

Cookies consist mainly of flour, sugar, and fat, and therefore the addition of pulse flour could improve their nutritional profile. Researchers have published several studies that show how it is possible to reformulate cookies with the addition of different types of pulses such as chickpeas (up to 10%) [70], lupins (up to 20%) [70], green lentils (from 25 to 100%) [71], and navy beans (up to 30%) [72]. These studies showed that, in cookies, the protein content increased proportionally with the addition of pulse flour while reducing dough spread. Pulse flour incorporation leads to darker surface color and a proportional increase in the hardness of the product. Using pulses to enrich bakery products is particularly suitable for gluten-free formulations, in fact, gluten does not play a key role in cookie-making. Malcomson et al. [73] showed that adding 20% yellow pea flour to gluten-free raw materials such as rice flour and tapioca starch did not modify the characteristics of cookies in terms of acceptability and texture.

In cake-baking, several types of pulse flours can be used as an ingredient, such as chickpeas [74] or peas [75]. Gomez et al. [75] focused on cake volume and observed a substantial decrease in the sample supplemented with pulse proteins; also, the resultant bubbles were smaller and more uniformly distributed. Firmness increased while springiness and cohesiveness decreased.

Another product that is well suited for reformulation with pulse flour are crackers. Malcomson et al. [73] added 30% of whole green lentil flour to a commercial cracker formulation. His findings show that crackers supplemented with lentil flour results in a protein-rich cracker with twice the total dietary fiber of wheat crackers. Crackers with lentil flour were darker in color, but their crisp texture and peppery flavor were considered acceptable and comparable to the control. Considering the non-essential role of gluten, crackers are suited not only for reformulations but also complete replacement with pulses including chickpea, green, red lentils pinto bean, navy bean, and yellow pea flours. The pulse-based, gluten-free cracker products investigated by Han et al. [76] have proved to be appealing for consumers, thanks to their health benefits. The sensory aspects of this cracker in terms of color, texture and taste were judged positively and were comparable with existing products on the market.

4. Main Barriers to the Use of Pulses and Potential Solutions

Using pulses in food formulations presents some challenges that need to be solved, in view of the nutritional benefits related to their consumption. The first difficulty to be faced is the presence of antinutritional factors, mainly phytic acid and tannins, in the seeds [77], which results in bloating and vomiting after ingestion of raw pulse seeds or flour [78,79]. Nevertheless, the anti-nutritional components may be reduced using different methods such as those recently reviewed by Patterson et al. [80]. The oldest and still widely used method to reduce antinutritional compounds consists of soaking, which leads to a reduction in phytate, which transfers to the soaking water [80].

Another method, dehulling consists of removing the outer layer of seed, which reduces cooking time, removes some antinutritional compounds (e.g., tannins) and improves protein digestibility [81]. Finally, thermal treatments, including extrusion, significantly decrease the presence of antinutritional compounds by eliminating heat-labile antinutrients [82]. Besides antinutritional factors, the sensory profile of pulses—i.e., their beany or bitter flavor profile—which depends on the type of pulses, greatly decrease their acceptability and thus their consumption. Traditionally, fermentation and germination have been used to enhance both the nutritional and sensory profiles of pulses, thanks to the production of aroma compounds and sugars [4].

Finally, incorporating pulses in cereal-based products causes important technological issues. In the case of bread, quality is related to an optimum balance of rheologically important gluten-forming proteins (i.e., gliadins and glutenins) and the addition of pulse flour to the wheat flour matrix leads to variations that inevitably worsen bread quality. The presence of pulse proteins not only dilutes gluten but also causes competition between wheat (gliadin and glutenin) and pulse (albumin and glogulin) proteins. Specifically, pulse proteins have a greater number of hydroxyl groups and for this reason they have a higher capacity for water binding [83]. Pulse fiber has also been reported to compromise gluten–gliadin strand formation [67,84]. Two main approaches can be taken to enhance the quality of final products. One depends on choosing suitable ingredients. The second approach consists of applying (bio)-technological treatments to the raw material.

4.1. Ingredients

Bread quality can be reestablished by using other ingredients, including vital gluten [68], hydrocolloids [64] or emulsifiers [85]. The fortification of flour with the addition of vital wheat gluten (0.1 g/gram flour) improves its rheology profile compared to that of the control blend, by increasing dough mixing stability, extensibility, and resistance to extension [68]. In particular, adding gluten to wheat–lentil composites significantly increases loaf volume for blends with <40% lentil concentration. Specifically, the concentration of gluten used by Portman et al. [68] could recover the possible loss of loaf volume caused by the addition of 5–15% lentil flour.

The combination of gluten (5%) and carboxymethylcellulose (5%) was effective in restoring or even improving the quality profile of breads formulated at maximum substitution levels of chickpea (20%), green pea (20%) and soybean (14%) flours [64].

Emulsifiers and pectin also significantly improved dough rheology, as well as the nutritional and sensory attributes of pulse-enriched bread [85]. The addition of emulsifiers (up to 1%) in a chickpea-enriched wheat bread significantly increased bread volume, while decreasing crumb firmness. The addition of emulsifiers can help to strengthen the gluten network, which in turn allows for greater pore size expansion resulting in a more porous bread crumb [69].

However, such ingredients in baked products may increase costs and might not satisfy consumer demand for clean label products. As an alternative, using a strong wheat flour could partially compensate for gluten dilution due to pulses. As mentioned above, pulses are generally incorporated in common wheat flours and the maximum enrichment level is 10%. However, when durum wheat semolina is used, the maximum enrichment level for yellow pea flour could go as high as 20%, producing a bread that was more appreciated and more similar to the control [86]. However, the decrease in volume and crumb porosity at 20–30% enrichment level might be counterbalanced by higher dietary fiber and lower glycemic index of the breads [86].

4.2. (Bio)-Technological Treatments

Besides the use of suitable ingredients, some biotechnological approaches—including air classification, fermentation, and germination—seem to be effective for enhancing the technological properties of pulses, their fractions, flours and/or enriched products. Table 3 summarizes the main results of the most recent scientific efforts in this field. Conflicting results among studies might be due to differences in plant species as well as variations in processing conditions.

Table 3. Aim and main results of the most recent studies on the (bio)-technological approaches applied to pulses.

Air Classification

Reference	Type of Pulses	Aim	Outcome
Rempel et al. [87]	pea	To assess the effects of milling and air classification on chemical composition.	Production of fine fractions having 90% of particles diameters smaller than 22 µm and high in protein (85–87%), fat (74–95%), and minerals (66–76%). Production of protein (yield: 20%, size: ≤15 µm) and starch (yield: 80%, size: 15–45 µm) fractions. The latter characterized by high viscosity and high resistant starch that might be used in food formulations to lower the glycemic index and/or increase viscosity of foods.
Simons et al. [88]	pinto bean	To produce high-starch fractions and assess their potential applications.	
Pelgrom et al. [89]	pea, lupine	To assess the effects of processing on the effectiveness of air classification.	Hydration, de-hulling or defatting prior to air classification were found effective in increasing protein yield and content.
Coda et al. [90]	faba bean	To enhance flour functionality by using fractions obtained by air classification.	Production of a starch-rich fraction with a low content in antinutritional factors.
Pelgrom et al. [91]	pea, bean, chickpea, lentil	To optimize the separation of starch granules from cell wall fibers and protein bodies.	Optimization of separation when the particle size distribution of flour overlaps with that of isolated starch granules.
Gómez et al. [75]	pea	To assess starch fraction suitability in cake making.	Using starch concentrate fraction did not affect negatively on cake quality but was found unacceptable for consumers. On the other hand, using protein fraction negatively affects cake quality.

Table 3. Cont.

Reference	Type of legumes	Aim	Outcome
Coda et al. [92]	faba bean	To investigate the effects of pulse sourdough on bread quality.	Using pulse sourdough positively affects the amino acid profile, protein digestibility, protein biological value, and glycemic index of bread.
Xu et al. [93]	faba bean	To assess the potential of different lactic acid bacteria in the production of exopolysaccharides and their impact on product texture.	Ln. pseudomesenteroides DSM 20193 showed the highest potential in the production of exopolysaccharides and texture modification in the related dough.
Rizzello et al. [94]	faba bean	To assess the effects of fermentation on of the pyrimidine glycoside vicine and convicine.	48 h of incubation with L. plantarum led to the degradation of the pyrimidine glycosides and aglycone derivatives.
Curiel et al. [95]	nineteen traditional Italian legumes	To assess the effects of sourdough fermentation on the functional and nutritional characteristics of pulses.	Fermentation promoted an increase in free amino acids, soluble fibers, and total phenols. Raffinose and condensed tannins decreased, while the level of gamma-aminobutyric acid, antioxidant and phytase activities markedly increased.
Rizzello et al. [96]	nineteen traditional Italian legumes	To investigate the effects of fermentation on the concentration of lunasin-like polypeptides.	Sourdough fermentation increased the amount of lunasin-like polypeptides, due to proteolysis of the native proteins. A marked inhibitory effect on the proliferation of Caco-2 cells was also observed.
Coda et al. [90]	faba bean	To assess the effects of air classification and lactic acid bacteria fermentation on the decrease in anti-nutritional factors and starch and protein digestibility of pulses.	The combination of air classification and fermentation was effective in decreasing/removing the anti-nutritional factors as well as improving the free amino acid content and protein digestibility.
Rizzello et al. [97]	chickpea, lentil, bean	To evaluate the effects of fermentation on nutritional, sensory and functional characteristics of pulse-enriched bread.	L. plantarum was the dominant lactic acid bacteria species in the wheat-legume sourdough. Using sourdough maximized the nutritional (by increasing the essential free amino acids, phenols and dietary fiber, and decreasing the hydrolysis index), sensory and functional properties of pulse-enriched bread.

Table 3. Cont.

Reference	Type of legumes	Aim	Outcome
Ouazib et al. [98]	chickpea	To investigate the impact of germination on rheological and bread-making performance of pulses.	Changes in starch upon germination significantly affected the rheological properties of the related flour. Germination negatively affected the overall acceptability of bread.
Ertaş [99]	lupin	To study the effects of sprouting on the physical and chemical properties of pulses and their bread-making performance	Sprouting of pulses enhanced the technological (volume, specific volume, symmetry and texture) and nutritional properties of bread.
Mondor et al. [100]	pea	To assess the effect of malted peas (10%) on bread quality.	The malting process did not affect the mixing property of the dough.
Marengo et al. [101]	chickpea	To assess the impact of sprouting on macromolecular and micronutrient profiles and rheological properties of chickpeas and chickpea flour–enriched dough (wheat/chickpea ratio = 100:20)	Sprouting enhanced the reticulating ability of proteins. Starch changes upon sprouting did not interfere with dough mixing properties and improved its leavening properties. Thus, sprouting of pulses might provide a good opportunity for developing new products with increased nutritional value.
Montemurro et al. [102]	chickpea	To investigate the effects of germination and sourdough fermentation on grain quality	Combining fermentation with sprouting further enhanced the nutritional and functional characteristics of flours, through the release of peptides and free amino acids, phenolic compounds and soluble fibers, and the decrease in several antinutritional factors. Bread enriched in fermented sprouted flour showed peculiar sensory profiles, and high protein digestibility and low starch availability, compared to the control sample.

4.3. Air Classification

As mentioned before, pulses are an interesting source of proteins and other nutrients. As regards protein isolation, applying wet fractionation under alkaline or acidic conditions yields relatively pure protein isolate, up to and sometimes exceeding 90%. However, the processing conditions (in terms of temperature and pH) are sometimes responsible for protein denaturation and loss of functionality. Moreover, wet extraction impacts negatively on the environment due to the amount of water and energy required. Consequently, dry separation processes are nowadays the preferred methods to separate plant proteins while maintaining their functionality. In this context, air classification has been widely studied, and its basic principles, together with recent applications have recently been reviewed [103,104]. Briefly, this technique exploits the centrifugal and gravitational forces induced by air to separate flour into fine and coarse particles differing in size and density. As a result, flours are separated into fractions characterized by different composition [105]. Feed particles must be sufficiently small and disaggregated in order for air classification to fractionate cell components [105]. For this reason, the efficiency of the separation is enhanced by the size reduction of the flour (by milling) prior to air classification. In addition, in starch-rich legumes, such as peas, starch granules (±20 µm) are embedded in a matrix of protein bodies (1–3 µm), which is fragmented during milling into particles smaller than the starch granules. Then, starch and proteins are easily separated based on size and/or density by air classification. Further information on milling of leguminous commodities has been recently reviewed by Thakur et al. [20]. Very fine milling is not optimal, because starch and fiber particles should be larger than the protein bodies [106]. On the contrary, coarse milling was found optimal to release protein bodies from their matrix, while further milling compromised the purity of the fine fraction [89,107].

Numerous studies dealt with the use of air classification up until the 1990s. To the best of our knowledge, most of these studies focused on the functional and nutritional features of these fractions, whereas studies about the incorporation of these flours in baked-products are scarce or null. More recently, Gómez et al. [75] highlighted the positive effect of using the starchy fraction from pea flour on the specific volume of sponge cakes. On the other hand, a worsening of the overall quality of the product was observed when fractions with high protein content were incorporated in the place of starch, which plays a structural role in cakes.

4.4. Fermentation

Sourdough fermentation, one of the oldest food biotechnologies, has been widely studied and recently rediscovered for its positive effects on the sensory, textural and nutritional features of baked products. Such changes are related to several biochemical modifications including acidification, proteolysis, activation of several enzymes, and synthesis of metabolites, which affect both dough and bread functionality [108]. More recently, sourdough fermentation has been proposed as a pre-treatment to stabilize or to increase the functional value of wheat-milling by-products [109–111]. Moreover, Gobbetti et al. [112] summarizes the most relevant effects of sourdough fermentation on legume flours, in terms of decreases in antinutritional factors (e.g., phytic acid concentration and trypsin inhibitory activity), and raffinose family oligosaccharides. On the other hand, fermentation promotes an increase in free amino acids, gamma-aminobutyric acid, soluble fibers, total phenol concentrations. Although several studies have focused on the nutritional benefits of pulse fermentation, few of them have investigated the effects of sourdough fermentation on dough rheology and the bread-making performance of pulse-enriched flours. Thus, further studies should be focused on such aspects in view of the potential use of sourdough fermentation as an interesting strategy to improve the sensory quality of pulse-enriched breads. Indeed, the sour flavor, which is typical of fermentation with lactic acid bacteria, might hide the off flavor of pulses and thus gain consumer acceptance of pulse-enriched breads [113].

4.5. Germination

Germination (also known as sprouting) is a re-emerging trend in healthy foods, thanks to its role in improving taste and nutritional properties. Germination is an easy process that traditionally takes place at home. It consists of soaking grains in water until they reach the moisture content necessary to start the growth of seedlings. After the soaking water is drained, the seeds are left to germinate. The germinated grains are then consumed in the form of sprouts or further processed (i.e., dried or roasted). Challenges regarding safety and reproducibility of process, must be solved for industrial applications and to guarantee a safe product with consistent features. Monitoring the process seems the only way to obtain a sprouted product with improved nutritional and sensory properties while maintaining flour performance, which ensures consistent product functionality [114].

Sprouting has been applied for millennia to pulses to reduce their anti-nutritional components, such as trypsin inhibitors and phytic acid. At the same time, enzymes produced during sprouting are also able to degrade ROFs (raffinose family of oligosaccharides) into shorter carbohydrates, eliminating some typical problems associated with their ingestion while developing sweet taste notes in germinated pulses. Moreover, sprouting influences the sensory properties of grains, giving them a typical flavor and odor generally perceived as pleasant.

The unique flavor profile of sprouted grains is due to the activation of endogenous amylolytic enzymes that transform complex starch molecules into simple oligosaccharides and sugars, which add natural sweetness to products. Thus, the sweetness of grain foods can be enhanced naturally by using the germination process. Moreover, during sprouting, reducing sugars and amino acids are released, which subsequently react during heating, giving rise to Maillard reaction products [115]. Finally, both germination and drying decrease musty and earthy odor notes, favoring the perception of roasted, nutty and intense flavor notes [116] and masking the unpleasant beany flavor in extruded products from soybean [117].

In addition to the nutritional and sensory aspects listed above, germination affects the technological performance of grains and related flours. Surprisingly, germination facilitates the dehulling process for brown chickpeas, mung peas, and pigeon peas [118]. Moreover, the process can influence the cooking properties of pulses by decreasing cooking times and reducing the amount of dispersed solids [119]. The decrease in cooking time for germinated grains is of great interest, since it would facilitate the preparation, and thus the consumption, of whole grains. Indeed, even while reducing the risk of cardiovascular disease and inflammation, the eating of grains is not so common in many countries, due to their long cooking times and bitter and pungent flavor notes [116].

The degree of germination-related effects greatly depends on the processing conditions used (i.e., time, temperature and relative humidity), since the biochemical events occurring during germination influence the quality of the ingredients. Recently, controlled germination has been carried out on peas and chickpeas [120]. The applied conditions (3 days, 22 °C and 90% relative humidity) induced mild structural modifications, sufficient to reduce anti-nutritional factors (e.g., phytic acid), without negatively affecting the nutritional quality of the grains (e.g., starch digestibility) [120]. The resulting flour has been proposed as an interesting ingredient for formulating enriched products [101]. Indeed, the reticulating ability of proteins improved as a result of sprouting, and the resultant starch did not interfere with dough development in formulations enriched with nutritionally significant levels of chickpea flour (i.e., 20%) [101]. Using germinated yellow peas in processed foods other than bread—such as white layer cakes, and extruded snacks—also resulted in end products with acceptable characteristics [121].

5. Conclusions

Today, consumers are more conscious about the environmental effects and nutritional benefits of foods. In this context, thanks to their nutritional and health-promoting properties, together with their low environmental impact, pulses can be considered a suitable raw material for food production. However, some factors limit their use on an industrial scale. Beside the presence of antinutritional factors, the sensory profile of pulses—i.e., their beany or bitter flavor profile—greatly decreases their acceptability. Traditionally, fermentation and germination have been used to enhance both the nutritional and sensory profiles of pulses, thanks to the production of aroma compounds and sugars. From a technological standpoint, incorporating pulses in cereal-based products is challenging due to the presence of fiber and non-gluten proteins. In wheat-based formulations, not only is gluten diluted by the presence of pulses proteins, but also wheat and pulse proteins compete for water and pulse proteins compete as rival water-binders. Two main approaches can be taken to enhance the quality of end-products. One is based on choosing suitable ingredients, such as strong wheat flour, vital gluten, hydrocolloids or emulsifiers. The second approach consists of the application of (bio)-technological treatments to the raw material, such as air classification, sourdough fermentation, and germination.

However, most studies seem to adopt an empiric approach from knowledge acquired from cereals, mainly varying ingredients and processing conditions rather than understanding the macromolecule organization associated with good or poor performance. Processes have the ability to modify the characteristics of macromolecules (mainly protein and starch) and their interactions. The extent of these changes is not only defined by the type of process but also by its intensity. In this context, efforts should be devoted to understanding the relation between types of pulses, extent of processes, biopolymer interactions, and product quality in terms of sensory, textural, and nutritional features. As regards the effects of processing on the nutritional characteristics of pulses, most studies often neglect the impact on dough rheology and bread quality. Thus, a multidisciplinary approach is recommended in order to provide solutions/strategies to satisfy both nutritional and technological demands.

Author Contributions: A.B. and A.M. were responsible for the bibliographic search and writing of the article. All of the authors read and approved the final manuscript.

Funding: This research received no external funding.

Conflicts of Interest: The authors declare no conflict of interest.

References

1. Tyler, R.; Wang, N.; Han, J. Composition, Nutritional Value, Functionality, Processing, and Novel Food Uses of Pulses and Pulse Ingredients. *Cereal Chem. J.* **2017**, *94*, 1. [CrossRef]
2. Rochfort, S.; Panozzo, J. Phytochemicals for Health, the Role of Pulses. *J. Agric. Food Chem.* **2007**, *55*, 7981–7994. [CrossRef] [PubMed]
3. Derbyshire, E. The Nutritional Value of Whole Pulses and Pulse Fractions. In *Pulse Foods: Processing, Quality and Nutraceutical Applications*; Tiwari, B.K., Gowen, A., McKenna, B., Eds.; Academic Press: London, UK, 2011; pp. 363–383.
4. Sozer, N.; Holopainen-Mantila, U.; Poutanen, K. Traditional and New Food Uses of Pulses. *Cereal Chem. J.* **2017**, *94*, 66–73. [CrossRef]
5. Hall, C.; Hillen, C.; Robinson, J.G. Composition, Nutritional Value, and Health Benefits of Pulses. *Cereal Chem. J.* **2017**, *94*, 11–31. [CrossRef]
6. Roland, W.S.U.; Pouvreau, L.; Curran, J.; Van De Velde, F.; De Kok, P.M.T. Flavor Aspects of Pulse Ingredients. *Cereal Chem. J.* **2017**, *94*, 58–65. [CrossRef]
7. Pratap, A.; Prajapati, U.; Singh, C.M.; Gupta, S.; Rathore, M.; Malviya, N.; Tomar, R.; Gupta, A.K.; Tripathi, S.; Singh, N.P. Potential, constraints and applications of in vitro methods in improving grain legumes. *Plant Breed.* **2018**, *137*, 235–249. [CrossRef]
8. Morris, J.B.; Wang, M.L. Updated review of potential medicinal genetic resources in the USDA, ARS, PGRCU industrial and legume crop germplasm collections. *Ind. Crop Prod.* **2018**, *123*, 470–479. [CrossRef]
9. Warsame, A.O.; O'Sullivan, D.M.; Tosi, P. Seed Storage Proteins of Faba Bean (Vicia faba L): Current Status and Prospects for Genetic Improvement. *J. Agric. Food Chem.* **2018**, *66*, 12617–12626. [CrossRef]
10. Farooq, M.; Hussain, M.; Usman, M.; Farooq, S.; Alghamdi, S.S.; Siddique, K.H.M. Impact of Abiotic Stresses on Grain Composition and Quality in Food Legumes. *J. Agric. Food Chem.* **2018**, *66*, 8887–8897. [CrossRef]
11. Luna-Vital, D.; De Mejia, E.G. Peptides from legumes with antigastrointestinal cancer potential: Current evidence for their molecular mechanisms. *Curr. Opin. Food Sci.* **2018**, *20*, 13–18. [CrossRef]

12. Harouna, D.V.; Venkataramana, P.B.; Ndakidemi, P.A.; Matemu, A.O. Under-exploited wild Vigna species potentials in human and animal nutrition: A review. *Glob. Food Secur.* **2018**, *18*, 1–11. [CrossRef]
13. Harsha, P.S.C.S.; Wahab, R.A.; Aloy, M.G.; Madrid-Gambin, F.; Estruel-Amades, S.; Watzl, B.; Andrés-Lacueva, C.; Brennan, L. Biomarkers of legume intake in human intervention and observational studies: A systematic review. *Genes Nutr.* **2018**, *13*, 25. [CrossRef] [PubMed]
14. Awika, J.M.; Rose, D.J.; Simsek, S. Complementary effects of cereal and pulse polyphenols and dietary fiber on chronic inflammation and gut health. *Food Funct.* **2018**, *9*, 1389–1409. [CrossRef] [PubMed]
15. Chhikara, N.; Devi, H.R.; Jaglan, S.; Sharma, P.; Gupta, P.; Panghal, A. Bioactive compounds, food applications and health benefits of Parkia speciosa (stinky beans): A review. *Agric. Food Secur.* **2018**, *7*, 46. [CrossRef]
16. Yi-Shen, Z.; Shuai, S.; Fitzgerald, R. Mung bean proteins and peptides: Nutritional, functional and bioactive properties. *Food Nutr. Res.* **2018**, *62*, 1290–1300. [CrossRef]
17. Cabanillas, B.; Jappe, U.; Novak, N. Allergy to peanut, soybean, and other legumes: Recent advances in allergen characterization, stability to processing and IgE cross-reactivity. *Mol. Nutr. Food Res.* **2018**, *62*, 1700446. [CrossRef] [PubMed]
18. Avilés-Gaxiola, S.; Chuck-Hernández, C.; Serna Saldivar, S.O. Inactivation methods of trypsin inhibitor in legumes: A review. *J. Food Sci.* **2018**, *83*, 17–29. [CrossRef] [PubMed]
19. Jeong, D.; Han, J.-A.; Liu, Q.; Chung, H.-J. Effect of processing, storage, and modification on in vitro starch digestion characteristics of food legumes: A review. *Food Hydrocoll.* **2019**, *90*, 367–376. [CrossRef]
20. Thakur, S.; Scanlon, M.G.; Tyler, R.T.; Milani, A.; Paliwal, J. Pulse Flour Characteristics from a Wheat Flour Miller's Perspective: A Comprehensive Review. *Compr. Rev. Food Sci. Food Saf.* **2019**, *18*, 775–797. [CrossRef]
21. Scanlon, M.G.; Thakur, S.; Tyler, R.T.; Milani, A.; Der, T.; Paliwal, J. The critical role of milling in pulse ingredient functionality. *Cereal Foods World* **2018**, *63*, 201–206.
22. Vishwakarma, R.K.; Shivhare, U.S.; Gupta, R.K.; Yadav, D.N.; Jaiswal, A.; Prasad, P. Status of pulse milling processes and technologies: A review. *Crit. Rev. Food Sci. Nut.* **2018**, *58*, 1615–1628. [CrossRef] [PubMed]
23. Van Der Poel, A. Effect of processing on antinutritional factors and protein nutritional value of dry beans (Phaseolus vulgaris L.). A review. *Anim. Feed. Sci. Technol.* **1990**, *29*, 179–208. [CrossRef]
24. Nkhata, S.G.; Ayua, E.; Kamau, E.H.; Shingiro, J.B. Fermentation and germination improve nutritional value of cereals and legumes through activation of endogenous enzymes. *Food Sci. Nutr.* **2018**, *6*, 2446–2458. [CrossRef] [PubMed]
25. Boukid, F.; Zannini, E.; Carini, E.; Vittadini, E. Pulses for bread fortification: A necessity or a choice? *Trends Food Sci. Technol.* **2019**, *88*, 416–428. [CrossRef]
26. Rehman, H.M.; Cooper, J.W.; Lam, H.M.; Yang, S.H. Legume biofortification is an underexploited strategy for combatting hidden hunger. *Plant Cell Environ.* **2019**, *42*, 52–70. [CrossRef] [PubMed]
27. Zhong, L.; Fang, Z.; Wahlqvist, M.L.; Wu, G.; Hodgson, J.M.; Johnson, S.K. Seed coats of pulses as a food ingredient: Characterization, processing, and applications. *Trends Food Sci. Technol.* **2018**, *80*, 35–42. [CrossRef]
28. Foschia, M.; Horstmann, S.W.; Arendt, E.K.; Zannini, E. Legumes as Functional Ingredients in Gluten-Free Bakery and Pasta Products. *Annu. Rev. Food Sci. Technol.* **2017**, *8*, 75–96. [CrossRef] [PubMed]
29. Jarpa-Parra, M. Lentil protein: A review of functional properties and food application. An overview of lentil protein functionality. *Int. J. Food Sci. Technol.* **2018**, *53*, 892–903. [CrossRef]
30. Burger, T.G.; Zhang, Y. Recent progress in the utilization of pea protein as an emulsifier for food applications. *Trends Food Sci. Technol.* **2019**, *86*, 25–33. [CrossRef]
31. Sharif, H.R.; Williams, P.A.; Sharif, M.K.; Abbas, S.; Majeed, H.; Masamba, K.G.; Safdar, W.; Zhong, F. Current progress in the utilization of native and modified legume proteins as emulsifiers and encapsulants—A review. *Food Hydrocoll.* **2018**, *76*, 2–16. [CrossRef]
32. Shevkani, K.; Singh, N.; Chen, Y.; Kaur, A.; Yu, L. Pulse proteins: Secondary structure, functionality and applications. *J. Food Sci. Technol.* **2019**, *56*, 2787–2798. [CrossRef] [PubMed]
33. Lam, A.C.Y.; Can Karaca, A.; Tyler, R.T.; Nickerson, M.T. Pea protein isolates: Structure, extraction, and functionality. *Food Rev. Int.* **2018**, *34*, 126–147. [CrossRef]
34. Best, D. 10 things to know about pulses. *Cereal Foods World* **2010**, *58*, 105. [CrossRef]
35. Maikhuri, R.; Dangwal, D.; Negi, V.S.; Rawat, L. Evaluation of symbiotic nitrogen fixing ability of legume crops in Central Himalaya, India. *Rhizosphere* **2016**, *1*, 26–28. [CrossRef]

36. Crews, T.; Peoples, M. Legume versus fertilizer sources of nitrogen: Ecological tradeoffs and human needs. *Agric. Ecosyst. Environ.* **2004**, *102*, 279–297. [CrossRef]
37. Mekonnen, M.M.; Hoekstra, A.Y. A global and high-resolution assessment of the green, blue and grey water footprint of wheat. *Hydrol. Earth Syst. Sci. Discuss.* **2010**, *7*, 2499–2542. [CrossRef]
38. Tiwari, B.; Singh, N. Major Constituents of Pulses. In *Pulse Chemistry and Technology*; Tiwari, B., Singh, N., Eds.; RSC Publishing: Cambridge, UK, 2012; pp. 34–51.
39. Tester, R.F.; Karkalas, J.; Qi, X. Starch—Composition, fine structure and architecture. *J. Cereal Sci.* **2004**, *39*, 151–165. [CrossRef]
40. Chibbar, R.N.; Ambigaipalan, P.; Hoover, R. REVIEW: Molecular Diversity in Pulse Seed Starch and Complex Carbohydrates and Its Role in Human Nutrition and Health. *Cereal Chem. J.* **2010**, *87*, 342–352. [CrossRef]
41. Brummer, Y.; Kaviani, M.; Tosh, S.M. Structural and functional characteristics of dietary fibre in beans, lentils, peas and chickpeas. *Food Res. Int.* **2015**, *67*, 117–125. [CrossRef]
42. Tosh, S.M.; Yada, S. Dietary fibres in pulse seeds and fractions: Characterization, functional attributes, and applications. *Food Res. Int.* **2010**, *43*, 450–460. [CrossRef]
43. Boye, J.; Zare, F.; Pletch, A. Pulse proteins: Processing, characterization, functional properties and applications in food and feed. *Food Res. Int.* **2010**, *43*, 414–431. [CrossRef]
44. Tiwari, B.K.; Singh, N. *Pulse Chemistry and Technology*, 1st ed.; Royal Society of Chemistry: London, UK, 2012; pp. 30–35.
45. Wang, S.; Melnyk, J.P.; Tsao, R.; Marcone, M.F. How natural dietary antioxidants in fruits, vegetables and legumes promote vascular health. *Food Res. Int.* **2011**, *44*, 14–22. [CrossRef]
46. Messina, V. Nutritional and health benefits of dried beans. *Am. J. Clin. Nutr.* **2014**, *100*, 437S–442S. [CrossRef]
47. Wallace, T.C.; Murray, R.; Zelman, K.M. The Nutritional Value and Health Benefits of Chickpeas and Hummus. *Nutrients* **2016**, *8*, 766. [CrossRef] [PubMed]
48. Hanson, M.G.; Zahradka, P.; Taylor, C.G. Lentil-based diets attenuate hypertension and large-artery remodelling in spontaneously hypertensive rats. *Br. J. Nutr.* **2014**, *111*, 690–698. [CrossRef]
49. Dahl, W.J.; Foster, L.M.; Tyler, R.T. Review of the health benefits of peas (Pisum sativum L.). *Br. J. Nutr.* **2012**, *108*, S3–S10. [CrossRef] [PubMed]
50. Mudryj, A.N.; Yu, N.; Aukema, H.M. Nutritional and health benefits of pulses. *Appl. Physiol. Nutr. Metab.* **2014**, *39*, 1197–1204. [CrossRef]
51. Havemeier, S.; Erickson, J.; Slavin, J. Dietary guidance for pulses: The challenge and opportunity to be part of both the vegetable and protein food groups. *Ann. N. Y. Acad. Sci.* **2017**, *1392*, 58–66. [CrossRef]
52. Patterson, C.A.; Maskus, H.; Dupasquier, C. Pulse crops for health. *Cereal Foods World* **2009**, *54*, 108–112. [CrossRef]
53. McCrory, M.A.; Hamaker, B.R.; Lovejoy, J.C.; Eichelsdoerfer, P.E. Pulse Consumption, Satiety, and Weight Management1. *Adv. Nutr.* **2010**, *1*, 17–30. [CrossRef]
54. Blackburn, G. Effect of Degree of Weight Loss on Health Benefits. *Obes. Res.* **1995**, *3*, 211s–216s. [CrossRef] [PubMed]
55. Bazzano, L.A.; Thompson, A.M.; Tees, M.T.; Nguyen, C.H.; Winham, D.M. Non-soy legume consumption lowers cholesterol levels: A meta-analysis of randomized controlled trials. *Nutr. Metab. Cardiovasc. Dis.* **2011**, *21*, 94–103. [CrossRef] [PubMed]
56. Jenkins, D.J.A.; Kendall, C.W.C.; Augustin, L.S.A.; Mitchell, S.; Sahye-Pudaruth, S.; Mejia, S.B.; Chiavaroli, L.; Mirrahimi, A.; Ireland, C.; Bashyam, B.; et al. Effect of Legumes as Part of a Low Glycemic Index Diet on Glycemic Control and Cardiovascular Risk Factors in Type 2 Diabetes Mellitus. *Arch. Intern. Med.* **2012**, *172*, 1653–1660. [CrossRef] [PubMed]
57. Jayalath, V.H.; De Souza, R.J.; Sievenpiper, J.L.; Ha, V.; Chiavaroli, L.; Mirrahimi, A.; Di Buono, M.; Bernstein, A.M.; Leiter, L.A.; Kris-Etherton, P.M.; et al. Effect of Dietary Pulses on Blood Pressure: A Systematic Review and Meta-analysis of Controlled Feeding Trials. *Am. J. Hypertens.* **2013**, *27*, 56–64. [CrossRef] [PubMed]
58. Mathers, J.C. Pulses and carcinogenesis: Potential for the prevention of colon, breast and other cancers. *Br. J. Nutr.* **2002**, *88*, 273–279. [CrossRef] [PubMed]
59. Roy, F.; Boye, J.; Simpson, B. Bioactive proteins and peptides in pulse crops: Pea, chickpea and lentil. *Food Res. Int.* **2010**, *43*, 432–442. [CrossRef]

60. Dewettinck, K.; Van Bockstaele, F.; Kühne, B.; Van de Walle, D.; Courtens, T.M.; Gellynck, X. Nutritional value of bread: Influence of processing, food interaction and consumer perception. *J. Cereal Sci.* **2008**, *48*, 243–257. [CrossRef]
61. Asif, M.; Rooney, L.W.; Ali, R.; Riaz, M.N. Application and Opportunities of Pulses in Food System: A Review. *Crit. Rev. Food Sci. Nutr.* **2013**, *53*, 1168–1179. [CrossRef]
62. Aïder, M.; Sirois-Gosselin, M.; Boye, J.I. Pea, Lentil and Chickpea Protein Application in Bread Making. *J. Food Res.* **2012**, *1*, 160–173. [CrossRef]
63. Kohajdová, Z.; Karovičová, J.; Magala, M. Effect of lentil and bean flours on rheological and baking properties of wheat dough. *Chem. Pap.* **2013**, *67*, 398–407. [CrossRef]
64. Angioloni, A.; Collar, C. High legume-wheat matrices: An alternative to promote bread nutritional value meeting dough viscoelastic restrictions. *Eur. Food Res. Technol.* **2012**, *234*, 273–284. [CrossRef]
65. Dabija, A.; Codină, G.G.; Fradinho, P. Effect of yellow pea flour addition on wheat flour dough and bread quality. *Rom. Biotech. Lett.* **2017**, *22*, 12888–12897.
66. Mohammed, I.; Ahmed, A.R.; Senge, B. Dough rheology and bread quality of wheat–chickpea flour blends. *Ind. Crop Prod.* **2012**, *36*, 196–202. [CrossRef]
67. Dalgetty, D.D.; Baik, B.K. Fortification of Bread with Hulls and Cotyledon Fibers Isolated from Peas, Lentils, and Chickpeas. *Cereal Chem. J.* **2006**, *83*, 269–274. [CrossRef]
68. Portman, D.; Blanchard, C.; Maharjan, P.; McDonald, L.S.; Mawson, J.; Naiker, M.; Panozzo, J.F. Blending studies using wheat and lentil cotyledon flour-Effects on rheology and bread quality. *Cereal Chem. J.* **2018**, *95*, 849–860. [CrossRef]
69. Yamsaengsung, R.; Schoenlechner, R.; Berghofer, E. The effects of chickpea on the functional properties of white and whole wheat bread. *Int. J. Food Sci. Technol.* **2010**, *45*, 610–620. [CrossRef]
70. Hegazy, N.A.; Faheid, S. Rheological and sensory characteristics of doughs and cookies based on wheat, soybean, chickpea and lupine flour. *Food Nahrung.* **1990**, *34*, 835–841. [CrossRef]
71. Zucco, F.; Borsuk, Y.; Arntfield, S.D. Physical and nutritional evaluation of wheat cookies supplemented with pulse flours of different particle sizes. *LWT FOOD SCI. TECHNOL.* **2011**, *44*, 2070–2076. [CrossRef]
72. Hoojjat, P.; Zabik, M.E. Sugar-snap cookies prepared with wheat-navy bean-sesame seed flour blends. *Cereal Chem.* **1984**, *61*, 41–44.
73. Malcolmson, L.; Boux, G.; Bellido, A.-S.; Fröhlich, P. Use of Pulse Ingredients to Develop Healthier Baked Products. *Cereal Foods World* **2013**, *58*, 27–32. [CrossRef]
74. Gómez, M.; Oliete, B.; Rosell, C.M.; Pando, V.; Fernández, E. Studies on cake quality made of wheat–chickpea flour blends. *LWT Food Sci. Technol.* **2008**, *41*, 1701–1709. [CrossRef]
75. Gómez, M.; Doyagüe, M.J.; De La Hera, E. Addition of pin-milled pea flour and air-classified fractions in layer and sponge cakes. *LWT Food Sci. Technol.* **2012**, *46*, 142–147. [CrossRef]
76. Han, J.; Janz, J.A.; Gerlat, M. Development of gluten-free cracker snacks using pulse flours and fractions. *Food Res. Int.* **2010**, *43*, 627–633. [CrossRef]
77. Gilani, G.S.; Xiao, C.W.; Cockell, K.A. Impact of Antinutritional Factors in Food Proteins on the Digestibility of Protein and the Bioavailability of Amino Acids and on Protein Quality. *Br. J. Nutr.* **2012**, *108*, S315–S332. [CrossRef] [PubMed]
78. Campos-Vega, R.; Loarca-Piña, G.; Oomah, B.D. Minor components of pulses and their potential impact on human health. *Food Res. Int.* **2010**, *43*, 461–482. [CrossRef]
79. Lajolo, F.M.; Genovese, M.I. Nutritional Significance of Lectins and Enzyme Inhibitors from Legumes. *J. Agric. Food Chem.* **2002**, *50*, 6592–6598. [CrossRef] [PubMed]
80. Patterson, C.A.; Curran, J.; Der, T. Effect of Processing on Antinutrient Compounds in Pulses. *Cereal Chem. J.* **2017**, *94*, 2–10. [CrossRef]
81. Wood, J.A.; Malcolmson, L.J. Pulse Milling Technologies. In *Pulse Foods: Processing, Quality and Nutraceutical Applications*; Tiwari, B.K., Gowen, A., McKenna, B., Eds.; Academic Press: London, UK, 2011; pp. 193–221.
82. Tiwari, B.; Singh, N. Pulse Products and Utilisation. In *Pulse Chemistry and Technology*; Tiwari, B., Singh, N., Eds.; RSC Publishing: Cambridge, UK, 2012; pp. 254–279.
83. Turfani, V.; Narducci, V.; Durazzo, A.; Galli, V.; Carcea, M. Technological, nutritional and functional properties of wheat bread enriched with lentil or carob flours. *LWT Food Sci. Technol.* **2017**, *78*, 361–366. [CrossRef]
84. Wang, J.; Rosell, C.M.; De Barber, C.B. Effect of the addition of different fibres on wheat dough performance and bread quality. *Food Chem.* **2002**, *79*, 221–226. [CrossRef]

85. Ajibade, B.O.; Ijabadeniyi, O.A. Effects of pectin and emulsifiers on the physical and nutritional qualities and consumer acceptability of wheat composite dough and bread. *J. Food Sci. Technol.* **2019**, *56*, 83–92. [CrossRef]
86. Ficco, D.B.M.; Muccilli, S.; Padalino, L.; Giannone, V.; Lecce, L.; Giovanniello, V.; Del Nobile, M.A.; De Vita, P.; Spina, A. Durum wheat breads 'high in fibre' and with reduced in vitro glycaemic response obtained by partial semolina replacement with minor cereals and pulses. *J. Food Sci. Technol.* **2018**, *55*, 4458–4467. [CrossRef] [PubMed]
87. Rempel, C.; Geng, X.; Zhang, Y. Industrial scale preparation of pea flour fractions with enhanced nutritive composition by dry fractionation. *Food Chem.* **2019**, *276*, 119–128. [CrossRef] [PubMed]
88. Simons, C.; Hall III, C.; Biswas, A. Characterization of pinto bean high-starch fraction after air classification and extrusion. *J. Food Process. Pres.* **2017**, *41*, e13254. [CrossRef]
89. Pelgrom, P.J.; Wang, J.; Boom, R.M.; Schutyser, M.A. Pre- and post-treatment enhance the protein enrichment from milling and air classification of legumes. *J. Food Eng.* **2015**, *155*, 53–61. [CrossRef]
90. Coda, R.; Melama, L.; Rizzello, C.G.; Curiel, J.A.; Sibakov, J.; Holopainen, U.; Pulkkinen, M.; Sozer, N. Effect of air classification and fermentation by Lactobacillus plantarum VTT E-133328 on faba bean (Vicia faba L.) flour nutritional properties. *Int. J. Food Microbiol.* **2015**, *193*, 34–42. [CrossRef]
91. Pelgrom, P.J.M.; Boom, R.M.; Schutyser, M.A.I. Method Development to Increase Protein Enrichment during dry fractionation of starch-rich legumes. *Food Bioprocess Technol.* **2015**, *8*, 1495–1502. [CrossRef]
92. Coda, R.; Varis, J.; Verni, M.; Rizzello, C.G.; Katina, K. Improvement of the protein quality of wheat bread through faba bean sourdough addition. *LWT Food Sci. Technol.* **2017**, *82*, 296–302. [CrossRef]
93. Xu, Y.; Wang, Y.; Coda, R.; Säde, E.; Tuomainen, P.; Tenkanen, M.; Katina, K. In situ synthesis of exopolysaccharides by Leuconostoc spp. and Weissella spp. and their rheological impacts in fava bean flour. *Int. J. Food Microbiol.* **2017**, *248*, 63–71. [CrossRef]
94. Rizzello, C.G.; Losito, I.; Facchini, L.; Katina, K.; Palmisano, F.; Gobbetti, M.; Coda, R. Degradation of vicine, convicine and their aglycones during fermentation of faba bean flour. *Sci. Rep.* **2016**, *6*, 32452. [CrossRef]
95. Curiel, J.A.; Coda, R.; Centomani, I.; Summo, C.; Gobbetti, M.; Rizzello, C.G. Exploitation of the nutritional and functional characteristics of traditional Italian legumes: The potential of sourdough fermentation. *Int. J. Food Microbiol.* **2015**, *196*, 51–61. [CrossRef]
96. Rizzello, C.G.; Hernández-Ledesma, B.; Fernández-Tomé, S.; Curiel, J.A.; Pinto, D.; Marzani, B.; Coda, R.; Gobbetti, M. Italian legumes: Effect of sourdough fermentation on lunasin-like polypeptides. *Microb. Cell Factories* **2015**, *14*, 168. [CrossRef] [PubMed]
97. Rizzello, C.G.; Calasso, M.; Campanella, D.; De Angelis, M.; Gobbetti, M. Use of sourdough fermentation and mixture of wheat, chickpea, lentil and bean flours for enhancing the nutritional, texture and sensory characteristics of white bread. *Int. J. Food Microbiol.* **2014**, *180*, 78–87. [CrossRef] [PubMed]
98. Ouazib, M.; Garzon, R.; Zaidi, F.; Rosell, C.M. Germinated, toasted and cooked chickpea as ingredients for bread making. *J. Food Sci. Technol.* **2016**, *53*, 2664–2672. [CrossRef] [PubMed]
99. Ertaş, N. Technological and chemical characteristics of breads made with lupin sprouts. *Qual. Assur. Saf. Crops* **2014**, *7*, 313–319. [CrossRef]
100. Mondor, M.; Guévremont, E.; Villeneuve, S. Processing, characterization and bread-making potential of malted yellow peas. *Food Biosci.* **2014**, *7*, 11–18. [CrossRef]
101. Marengo, M.; Carpen, A.; Bonomi, F.; Casiraghi, M.C.; Meroni, E.; Quaglia, L.; Iametti, S.; Pagani, M.A.; Marti, A. Macromolecular and Micronutrient Profiles of Sprouted Chickpeas to Be Used for Integrating Cereal-Based Food. *Cereal Chem. J.* **2017**, *94*, 82–88. [CrossRef]
102. Montemurro, M.; Pontonio, E.; Gobbetti, M.; Rizzello, C.G. Investigation of the nutritional, functional and technological effects of the sourdough fermentation of sprouted flours. *Int. J. Food Microbiol.* **2019**, *302*, 47–58. [CrossRef] [PubMed]
103. Assatory, A.; Vitelli, M.; Rajabzadeh, A.R.; Legge, R.L. Dry fractionation methods for plant protein, starch and fiber enrichment: A review. *Trends Food Sci. Technol.* **2019**, *86*, 340–351. [CrossRef]
104. Schutyser, M.; Pelgrom, P.; Van Der Goot, A.; Boom, R. Dry fractionation for sustainable production of functional legume protein concentrates. *Trends Food Sci. Technol.* **2015**, *45*, 327–335. [CrossRef]
105. Andersson, A.A.M.; Andersson, R.; Åman, P. Air Classification of Barley Flours. *Cereal Chem. J.* **2000**, *77*, 463–467. [CrossRef]
106. Pelgrom, P.J.; Vissers, A.M.; Boom, R.M.; Schutyser, M.A. Dry fractionation for production of functional pea protein concentrates. *Food Res. Int.* **2013**, *53*, 232–239. [CrossRef]

107. Pelgrom, P.J.; Berghout, J.A.; Van Der Goot, A.J.; Boom, R.M.; Schutyser, M.A. Preparation of functional lupine protein fractions by dry separation. *LWT Food Sci. Technol.* **2014**, *59*, 680–688. [CrossRef]
108. Gobbetti, M.; Rizzello, C.G.; Di Cagno, R.; De Angelis, M. How the sourdough may affect the functional features of leavened baked goods. *Food Microbiol.* **2014**, *37*, 30–40. [CrossRef] [PubMed]
109. Marti, A.; Torri, L.; Casiraghi, M.C.; Franzetti, L.; Limbo, S.; Morandin, F.; Quaglia, L.; Pagani, M.A. Wheat germ stabilization by heat-treatment or sourdough fermentation: Effects on dough rheology and bread properties. *LWT Food Sci. Technol.* **2014**, *59*, 1100–1106. [CrossRef]
110. Rizzello, C.G.; Cassone, A.; Coda, R.; Gobbetti, M. Antifungal activity of sourdough fermented wheat germ used as an ingredient for bread making. *Food Chem.* **2011**, *127*, 952–959. [CrossRef]
111. Rizzello, C.G.; Coda, R.; Mazzacane, F.; Minervini, D.; Gobbetti, M. Micronized by-products from debranned durum wheat and sourdough fermentation enhanced the nutritional, textural and sensory features of bread. *Food Res. Int.* **2012**, *46*, 304–313. [CrossRef]
112. Gobbetti, M.; De Angelis, M.; Di Cagno, R.; Calasso, M.; Archetti, G.; Rizzello, C.G. Novel insights on the functional/nutritional features of the sourdough fermentation. *Int. J. Food Microbiol.* **2019**, *302*, 103–113. [CrossRef]
113. Pealoza-, J.; Mora-Escobedo, R.; Chanona-Prez, J.; Farrera-Rebollo, R.; Caldern-Domnguez, G.; De La Rosa-Angulo, G.J. *Sourdough and Bread Properties as Affected by Soybean Protein Addition*; IntechOpen: Rijeka, Croatia, 2011; pp. 387–402.
114. Marti, A.; Cardone, G.; Pagani, M.A.; Casiraghi, M.C. Flour from sprouted wheat as a new ingredient in bread-making. *LWT Food Sci. Technol.* **2018**, *89*, 237–243. [CrossRef]
115. Goesaert, H.; Slade, L.; Levine, H.; Delcour, J.A. Amylases and bread firming—An integrated view. *J. Cereal Sci.* **2009**, *50*, 345–352. [CrossRef]
116. Heiniö, R.; Noort, M.; Katina, K.; Alam, S.; Sozer, N.; De Kock, H.; Hersleth, M.; Poutanen, K. Sensory characteristics of wholegrain and bran-rich cereal foods—A review. *Trends Food Sci. Technol.* **2016**, *47*, 25–38. [CrossRef]
117. Obatolu, V.A. Nutrient and sensory qualities of extruded malted or unmalted millet/soybean mixture. *Food Chem.* **2002**, *76*, 129–133. [CrossRef]
118. Bellaio, S.; Zamprogna Rosenfeld, E.; Mane, D.; Jacobs, M. Novel process based on partial germination to enhance milling yield and nutritional properties of pulses. *Cereal Foods World* **2011**, *56*, A30.
119. Zamprogna Rosenfeld, E.; Bellaio, S.; Mane, D.; Jacobs, M. A new family of healthy, safe, and convenient food products based on partial germination of pulses. *Cereal Foods World* **2011**, *56*, A71.
120. Erba, D.; Angelino, D.; Marti, A.; Manini, F.; Faoro, F.; Morreale, F.; Pellegrini, N.; Casiraghi, M.C. Effect of sprouting on nutritional quality of pulses. *Int. J. Food Sci. Nutr.* **2019**, *70*, 30–40. [CrossRef] [PubMed]
121. Han, J.; Buchko, A. Development of a partial germination process for yellow peas and resultant pea flours in white layer cakes. *Cereal Foods World* **2014**, *59*, A42.

© 2019 by the authors. Licensee MDPI, Basel, Switzerland. This article is an open access article distributed under the terms and conditions of the Creative Commons Attribution (CC BY) license (http://creativecommons.org/licenses/by/4.0/).

Review

Bread Enrichment with Oilseeds. A Review

Beatriz de Lamo and Manuel Gómez *

Food Technology Area, College of Agricultural Engineering, University of Valladolid, 34004 Palencia, Spain; beatriz.de.lamo.santamaria@gmail.com
* Correspondence: pallares@iaf.uva.es; Tel.: +34-979108495

Received: 30 October 2018; Accepted: 19 November 2018; Published: 20 November 2018

Abstract: The use of oilseeds in bakery products has gained popularity in recent years, both for their organoleptic and nutritional characteristics. The aim of this work is to provide an overview of the studies centered on the use of oilseeds (flaxseed, chia, sunflower, pumpkin, sesame and poppyseed) in breads and other bakery products. This review highlights the effect of oilseeds on the mechanical and physical properties of bread according to the enrichment level, origin and way of addition (whole, crushed, oil or mucilage). In general, the incorporation of oilseeds improves the nutritional profile of bakery products with and without gluten, and provides several health benefits. Mucilages of oilseeds can also act as a fat replacer thanks to their properties. The incorporation of oilseeds modifies the rheology of the doughs, the volume of the products and their texture, affecting their organoleptic characteristics and their acceptability. Nevertheless, these changes will depend on the type of seed used, as well as on the method of addition.

Keywords: bread; fortification; sunflower; flaxseed; chia; sesame; mucilage

1. Introduction

Oilseeds are grains with a high amount of oil. They have been part of the human diet for a long time, but it has only been in the last few decades that their use and production have trended upward. The use of oilseeds in the modern market has increased due to the growing number of people concerned about a healthier lifestyle, as well as better knowledge of their attractive composition [1]. The chemical composition of oilseeds depends on their growing environment, genetics and processing conditions [2]. Generally, these seeds have lower carbohydrate content and higher protein content than cereals, high levels of fiber, and omega-6 and omega-3 essential fatty acids [3]. These seeds also enclose a high proportion of natural antioxidant compounds (tocopherol, beta-carotene chlorogenic acid, caffeic acid and flavonoids), vitamins and minerals [4–6]. Moreover, the absence of gluten in these seeds allows their ingestion by celiac disease sufferers [1,6]. Thanks to this composition, some of these seeds have proven effective in controlling and preventing metabolic diseases (hypertension, hypercholesterolemia, diabetes, coronary heart disease and several types of cancer) as well as providing interesting properties to foods (body, texture and taste improvement) [5–8].

Since bread is considered a staple food worldwide because of its nutritive value, low price and its simplicity of use, it is ideally suited for fortification with oilseeds. The impact of oilseeds on bread baking has been studied in many investigations. Oilseeds can be incorporated whole, crushed and as pressed oil, and their beneficial effect on health seems to depend on the method of addition [9]. Further, changes in dough rheology and the quality of final products, including their sensory acceptability, depends on whether they are incorporated whole or crushed. Other aspects, such as the oxidative stability of the enriched products, must be also considered [10]. In the case of chia and flax, the mucilaginous polysaccharide that these seeds exude when they are placed in aqueous medium can also be added to the formulation to provide advantageous technological properties in terms of food development. The hydration of grains or flours before their incorporation into the bread doughs, in order to extract these polysaccharides, can modify the rheology of the doughs and the quality of the breads [11]. These grains can also be added only on the surface of breads or other products, in which case the influence on the final product will be less. Differences in the development of breads have been reported based on the method of incorporation and the pretreatment of the oilseeds, among other things, and therefore there are still numerous challenges to solve in their use (oxidation of the lipids, the impact of the gluten network to the structure, etc.).

Despite the wide use of oilseeds in bakery products, scientific studies on them are scarce. The purpose of this review is to deepen the knowledge of the use of different oilseeds for the development of enriched breads, both from a nutritional and organoleptic point of view.

2. Nutritional Profile

Oilseeds, as their name suggests, have a high fat content that usually exceeds 40%, except for chia and pumpkin seeds, which have a lower percentage (Table 1). The low values of oil content in pumpkin seeds reported by the United States Department of Agriculture (USDA) may conflict with some varieties analysed by Stevenson et al. [8]. In their research, some varieties of pumpkin seeds were 30% lipids, while in the study by Seymen et al. [7] the oil contents of pumpkin seeds were between 33% and 47% depending on the variety. Furthermore, these fats stand out for their low level of saturated fatty acids and for their high content of oleic acid in sesame seeds; linoleic acid ($\omega 6$) in sunflower, sesame and poppy seeds; and linolenic acid ($\omega 3$) in flax and chia seeds. In the case of pumpkin seeds, apart from their high linoleic acid level, their oleic acid content is also highlighted [7,8,12]. These seeds also have significant protein content (between 15 and 20%) and a fiber percentage higher than that of cereal grains, with fiber content of flax and chia seeds over 25%. In the case of chia, different studies have reported higher values of fiber (between 35–40%) depending on the variety, and of which more than 85% is insoluble fiber [13–15].

Table 1. Nutrient content of some whole oilseeds (per 100g).

		Wheat	Sunflower	Flaxseed	Sesame	Chia	Pumpkin	Poppy
Nutrients	Water (g)	12.42	1.2	6.96	4.69	5.8	4.50	5.95
	Energy value (kcal)	332	582	534	573	486	446	525
	Protein (g)	9.61	19.33	18.29	17.73	16.54	18.55	17.99
	Total lipid (fat) (g)	1.95	49.8	42.16	49.67	30.74	19.40	41.56
	Carbohydrate (g)	74.48	24.07	28.88	23.45	42.12	53.75	28.13
	Fiber (g)	13.1	9	27.3	11.8	34.4	18.4	19.5
	Sugar (g)	1.02	2,73	1.55	0,30	N	N	2.99
Minerals	Calcium (mg)	33	70	255	975	631	55	1438
	Iron (mg)	3.71	3.8	5.73	14.55	7.72	3.31	9.76
	Magnesium (mg)	117	129	392	351	335	262	347
	Phosphorous (mg)	323	1115	642	629	860	92	870
	Potassium (mg)	394	850	813	468	407	919	719
	Sodium (mg)	3	655	30	11	16	18	26
	Zinc (mg)	2.96	5.29	4.34	7.75	4.58	10.30	7.9
Vitamins	Vitamin C (mg)	0	1.40	0.6	0	1.6	0.3	1
	Thiamin (mg)	0.297	0.106	1.644	0.791	0.62	0.034	0.854
	Riboflavin (mg)	0.188	0,246	0.161	0.247	0.17	0.052	0.1
	Niacin (mg)	5.347	7.04	3.08	4.515	8.83	0.286	0.896
	Vitamin B6 (mg)	0.191	0.804	0.473	0.79	N	0.037	0.247
	Folate (µg)	28	237	87	97	49	9	82
	Vitamin E (mg)	0.53	26.1	0,31	0,25	0,50	N	1.77

Table 1. Cont.

		Wheat	Sunflower	Flaxseed	Sesame	Chia	Pumpkin	Poppy
Lipids	Saturated (g)		5.219	3.663	6.957	3.33	3.670	4.517
	Monounsaturated (g)		9.505	7.527	18.759	2.309	6.032	5.982
	18:1 (g)		9.399	7.359	18.521	2.203	5.985	5.864
	Polyunsaturated (g)		32.884	28.73	21.773	23.665	8.844	28.569
	18:2 (g)		32.782	5.903	21.375	5.835	8.759	28.295
	18:3 (g)		0.069	22.813	0.376	17.830	0.077	0.0273

Source: United States Department of Agriculture (USDA) National Nutrient Database for Standard Reference (2018). N, there is no reliable information about the quantity of the nutrient.

Mineral content is noteworthy. The high calcium content of sesame, chia and poppyseed; the iron content of sesame and poppyseed; the zinc content of pumpkin; and the magnesium, phosphorus and potassium contents of most of these seeds stand out, and are higher than in wheat. As a negative aspect, sunflower seeds have high sodium content. They also stand out for their high level of vitamins B and E. In particular, sesame is notable for its high levels of vitamin B6 and folates; flaxseed for its high level of thiamine and folates; chia for its high level of niacin; and sunflower for its high content of niacin, vitamin B6, folates and vitamin E. Some of these seeds also show a high content of antioxidants. Thus, chia has a high concentration of polyphenols derived from caffeic acid and other antioxidant substances [13,14,16,17]. The antioxidant potential of flax [18], sesame [19,20], pumpkin [7] and sunflower seeds [21] has also been highlighted. In the case of chia and flax seeds, it has been demonstrated that the partial germination of the grains can increase both their antioxidant capacity and the content of phenolic compounds [22].

The interesting composition of these grains has attracted interest for their possible health benefits. Fiber consumption has been shown to be effective against CVD (Cardiovascular diseases), CVD mortality, coronary artery disease and several kinds of cancer [23], as well as against various gastrointestinal disorders, including: gastroesophageal reflux disease, duodenal ulcer, diverticulitis, constipation, and haemorrhoids [24]. The consumption of omega-3 fatty acids has also been associated with a reduction of cardiovascular morbidity and mortality, and dietary supplementation may also benefit patients with dyslipidaemia, atherosclerosis, hypertension, diabetes mellitus, metabolic syndrome, obesity, inflammatory diseases, neurological/neuropsychiatric disorders and eye diseases [25]. Lignans, a phytoestrogen that stands out among antioxidant substances, are found in important quantities in flax, sesame and pumpkin seeds and have shown promise in reducing growth of cancerous tumours, especially hormone-sensitive ones such as those of the breast, endometrium, and prostate [26]. Moreover, a synergistic effect between sesame lignans and the vitamin E activity of tocopherols has been proved. Vitamin E activity increases in the presence of these lignans, while the cholesterol-lowering effect of lignans is enhanced [27]. The stability of flax lignans to the conditions of baking processes and storage has also been demonstrated and thus no loss of functionality when incorporated into bakery products has been observed [28]. Likewise, it is known that chia helps to reduce postprandial blood glucose, the level of high-density lipoprotein in serum and diastolic blood pressure [29]. Moreover, different reviews suggest that the consumption of chia could help to increase the satiety index, prevent cardiovascular diseases, inflammatory and nervous system disorders, and diabetes, among others [6,30]. Flax seeds have shown a positive effect on cardiovascular diseases and hypercholesterolemic atherosclerosis [31] and the sesame consumption has been linked to positive effects against some illnesses such as cancer, oxidative stress, cardiovascular disease, osteoporosis and other degenerative diseases [32,33]. In the case of flaxseed, it has been demonstrated that the incorporation of the flour of this seed in a wide number of products does not affect the consumption preferences of the same by the population, so its use can be a good strategy to provide significant health-related benefits to patients with cardiovascular disease [34].

Many studies have focused on the nutritional enrichment of bakery products through the incorporation of this type of seed or by-products, such as oils, proteins or other derivatives. In general, an increase in the content of nutrients present in the seeds, or their derivatives, is confirmed, since there are no losses or significant changes in them during the baking process. As for the incorporation of proteins extracted from these seeds, Mansour et al. [35] observed that up to 22% replacement of wheat by pumpkin protein isolate is possible without adverse effects on bread quality. They also reported that the obtained breads have a higher level of protein, lysine and minerals, improving also the essential amino acid index. However, mineral enrichment was greater when the seed flour was incorporated. The study of El-Soukkary [36] confirmed these observations with isolated pumpkin seed protein and observed that the in vitro protein digestibility also improved. These effects are repeated with the incorporation of sesame proteins, but in this case, the maximum level of substitution was 18% [37]. The incorporation of sunflower seeds at levels of up to 16% allowed an increasing of the levels of tocopherol and certain minerals, such as Ca, Mg and Zn in normal and whole grain breads, as well as enriching them in fats with a high linoleic content [38]. The incorporation of chia, as seed or flour, increased the fiber content and omega-3 acids both in wheat breads and in gluten-free breads [39,40]. Some researchers have also studied the inclusion of chia flour up to 14% and have observed an increase in the levels of ash, protein, fiber and lipids [41]. Regarding the incorporation of flax, as seed or flour, the incorporation of milled brown flaxseed in breads at levels of up to 13% resulted in an increase in the content of fiber, P, K, Mg, Zn and linolenic (omega-3) [42]. These same authors observed a decrease in the total cholesterol, low-density lipoproteins (LDL) and the level of triglycerides in rats when they were fed with these breads compared to those fed with breads without addition of flaxseeds. On the other hand, the incorporation of ground flaxseed hulls at levels lower than 5% was sufficient to increase significantly the antioxidant activity in breads [43].

In general, the incorporation of these seeds, whole or ground, increases the content of fiber, proteins, minerals, omega-3, omega-6 and minerals. For this reason, the inclusion of oilseeds in gluten-free breads is of special interest, since it is known that the gluten-free diet can be low in fiber and minerals (iron, zinc, magnesium and calcium), and may contain excess saturated fats [44–47]. In fact, supplements of these micronutrients have been proposed to improve the quality of the gluten-free diet [48]. In general, gluten-free breads also have lower protein content and lower levels of certain minerals than wheat varieties [49–52]. Additionally, it must be considered that the incorporation of these seeds, together with ingredients such as milk powder in gluten-free breads, not only increases the mineral content of gluten-free breads, but can also increase their bioavailability, depending on the mineral and the starting flour [53]. In addition, several studies with rats have demonstrated the potential of gluten-free breads enriched with this type of seed to reduce the levels of triglyceride and improve the antiradical properties of serum [54].

Despite the clear evidence of the nutritional advantages of incorporating this type of seed in breads, the health benefits can sometimes depend on the method in which they are incorporated. Kuijsten et al. [9] showed that milling the flaxseeds improved the bioavailability of the enterolignans, and therefore their nutritional advantages increased. Similarly, the reduction of the particle size by wet micronisation increased the oral bioavailability of lignan glycosides from sesame meal [55]. However, when flaxseed products are incorporated into muffins or breads, the mammalian lignan production of flaxseed precursors depends on the time and dose, but not on the processing [56]. Although more research is needed in this regard, we must also consider the effects of this processing on the rheology of the doughs and the organoleptic quality of the breads. The stability of these products must also be considered during storage, something of great importance for industrial use. Although Malcolmson et al. [57] observed good stability of the milled flaxseed stored at room temperature in paper bags with plastic liners during a four-month period, other studies have shown that when the particle size decreases, the oxidative stability of these products decreases too [58]. Therefore, when milled oil seeds are used, aspects such as particle size and packaging conditions must be considered in order to maximise the stability of these products.

3. Flaxseed and Chia Seed Mucilages

Chia seeds and flaxseeds are characterised by being very rich in mucilage. The mucilage of flaxseeds, consisting mainly of water-soluble polysaccharides, has a great water absorption capacity and rheological properties similar to those of guar gum [59]. In fact, flaxseed mucilages have been proposed as a substitute for gluten in the preparation of gluten-free breads in order to improve their acceptability against a mixture of pectin and guar gum [60]. Its composition includes a neutral fraction, composed of arabinoxylans, and an acid fraction composed of polysaccharides similar to pectin [61]. As with flax mucilage, chia mucilages are extracted in water, more easily if the temperature is high; when the proportion of water-seed increases, the saline content of the water decreases and the pH increases [1], so the conditions of the mixture can affect the interactions with this component. In this research, the hydration manages to extract mucilage with yields of 7%, somewhat lower than the maximum achieved with flax, which was slightly higher than 9% [59]. The chia mucilage has also been compared with guar gum and locust bean gum because of its high hydration capacity [62] and great thickening power [63]. Its structure corresponds to that of arabinoxylans, but with greater amounts of uronic acids than those of flax, imparting anionic characteristics to the macromolecule [64]. On the other hand, its ability to increase the stability of oil-in-water (O/W) emulsions has been demonstrated [65]. In the case of flax mucilage, its good foam stability properties in aqueous solutions have also been demonstrated [66].

Just like other hydrocolloids, the use of chia mucilage as a stabiliser in ice cream [67] and as a substitute for fats in breads, cakes [68–70] or mayonnaise [71] has been proposed. The use of hydrocolloids with high water absorption capacity and thickening power, such as xanthan gum, in wheat bread increases the absorption of the doughs, the development time and stability in the kneading, generates more tenacious and less extensible doughs, and can increase the specific volume and the firmness of the breads [72]. Similarly, guar gum has proved efficient in improving the volume of breads made with non-frozen and frozen dough [73]. This type of hydrocolloid has also been proposed to reduce water losses in storage, both of finished breads [74] and doughs and breads under refrigeration [75]. Usually, gluten-free formulations have applied hydrocolloids to mimic the viscoelastic properties of gluten [76]. Among these hydrocolloids, those with high gas-holding capacity and high thickening power, such as guar gum or xanthan gum, have been suggested [77], and are some of the most widely used industrially. At a nutritional level, these gums are also interesting. Guar gum is capable of significantly decreasing the in vitro hydrolysis of starch, an effect that has been attributed both to guar galactomannan, which acts as a physical 'barrier' to alpha-amylase-starch interactions, and to the effect of guar gum on digesta viscosity [78]. Likewise, breads enriched with arabinoxylans can help control the glycaemic level in diabetics [79]. However, the incorporation of chia mucilage has only been able to reduce the glycaemic index estimated in the crust of pita breads, but to increase it in the crumb [80]. More research would be necessary but, in general, the use of chia and flax mucilage can be very interesting in baking, isolated or together with the flour or seeds, in which case its functionality will depend on hydration and its release from the original matrix to interact with the components of the dough.

4. Dough Rheology

The rheology of bread doughs gives us an idea of how they behave in different processes, and can also help us to predict some final characteristics of the breads. Current studies on the influence of oleaginous seeds and their flours on the rheology of bread doughs are limited, but we expect that there is an influence depending on their composition, since they are rich in oil, vegetable proteins and fiber. Thus, it is known that the addition of small amounts of oil or fat leads to greater flexibility and workability of the doughs, greater final volume, a finer alveolar structure, and a softer final texture [81]. The addition of fats and oils also delays starch retrogradation, prolonging the shelf-life of the breads. On the other hand, the vegetal proteins modify the farinographic properties in a different way according to their type [82,83], but in general they reduce the strength of the doughs and their extensibility. The incorporation of fiber usually increases water absorption, the mixing tolerance and the tenacity of the doughs, but reduces extensibility [84]. The type of incorporation must be also considered since, if oilseeds are incorporated as flour, their different components will interact with the components of the dough, will compete with them, or will establish synergies. However, if they are incorporated as seeds, although these seeds can also have an effect on the properties, their components will not interact since they will remain inside the seed coat.

In the farinographic analysis of doughs with chia at 5%, the incorporation of seeds does not modify the behaviour during the kneading but decreases absorption slightly [85]. However, in this same research the addition of whole or low-fat chia flour increased the stability of the doughs, and when defatted chia flour was added the absorption also increased. Contrary to what was observed in this study, Steffolani et al. [11] reported that a 5% addition of chia flour does not affect absorption, but it increases with a 15% addition. The incorporation of chia also reduced the stability of the doughs and increased the development time. The decrease in the stability of the doughs matches the observations of Koka and Anil [86] with the incorporation of flaxseed, Moreira et al. [87] with chia flour, and Matthews et al. [88] with the addition of different oilseed flours. A greater absorption of water was also observed with the incorporation of these flours in all these studies. In general, there seems to be agreement that the use of these types of flours does not modify the absorption or increase it, while the effect on stability is contradictory. The differences may be due to the different starting flours, the percentage and the type of oilseed flour used. In fact, the wheat flour used by Iglesias, Puig and Haros [85] presented lower stability values than those used by Steffolani et al. [11] or Koka and Anil [86], making it easier to increase this stability. Regarding the variety of seed used, Svec et al. [89] even observed that using different types of chia (white or black) modified the effect of these flours on the amylographic analysis, although, in both, the peak viscosity increased, the viscosity rise was faster and the pasting time decreased. These results coincide with what Verdu et al. [90] observed by Rapid Visco Analyser (RVA), which has been attributed to the presence of mucilage. The incorporation of flax and chia flours also tends to reduce the extensibility of the doughs and the energy of deformation, measured with the extensograph or with the alveograph [11,86], which can cause problems in the use of the doughs or in the retention of gas during fermentation. In fact, Steffolani et al. [11] observed that the gluten network of the doughs with 10% of seeds or chia flour was broken before the control dough, with gas escaping to the outside. This effect could be expected due to the dilution of the gluten located in the mixtures and responsible for these properties. Similar results have been observed due to the incorporation of lipids [91] and products with a high content of non-gluten-forming proteins [92], which can inhibit the formation of the gluten network.

In the case of gluten-free doughs, the use of equipment such as the alveograph or the extensograph is not possible because they are based on the measurement of the characteristics that the gluten confers on the doughs, such as extensibility or tenacity. It is usual to analyse the gluten-free doughs by conventional rheology, such as in oscillatory tests. There is limited research related to this, but Moreira et al. [87] observed that the incorporation of chia flour into doughs of chestnut flour reduced both the elastic and the viscous component. These results conflict with the observations of Zettel and Hitzmann [93] in doughs of sweet bread, but in this study the chia flour was used to replace fats together with a greater incorporation of water, so the changes in the rheology cannot be attributed exclusively to the incorporation of chia flour.

The previous hydration of chia flours or seeds can also have a great impact on the rheological properties of the doughs, since it allows the mucilage of this seed to flow into the water and interact better with the rest of the ingredients of the dough. Thus, the prehydration of the chia flour allows an increase of the absorption of doughs and reduces development times, although it does not modify either the alveographic properties [11] or the pasting properties of gluten-free flours [94].

In general, the modification of the rheology of the doughs with the incorporation of oilseeds or their flours can be confirmed. This modification seems to be more pronounced in the case of flours than of seeds although these changes are not very considerable if the amount of flour does not exceed

5%. The modification of the rheology of the doughs really depends on many factors, so one cannot draw general conclusions and each case must be studied.

5. Bread Quality

The effect of oilseed flours on breads is determined by their components. It is known that the incorporation of fibers [84], such as celluloses, or vegetable proteins [95] usually reduces the specific volume of the bread's original firmer textures. The effect of the incorporation of oils on breads depends on their added quantity. In small amounts and depending on the type of oil added the incorporation of oils can be beneficial, while in larger quantities they weaken the dough and have negative effects on the volume of the breads [96]. Therefore, depending on the added product (wholemeal, protein concentrate, defatted flour), its composition and the amount added, the effects on the quality of bread may vary. Generally, a negative effect on the volume of the breads is expected after incorporation. When non-ground whole seeds are used, a minor effect is expected since not all of their components interact with the wheat flour.

Chia flour is one of the most studied for baking processes. In general, wheat breads reduce their volume, increase hardness and present darker crumbs as the level of chia flour increases, as some research on bread fortification [11,97] or the substitution of saturated fats for chia flour [39] has shown. However, some studies have reported an increase in the volume and a reduction of the hardness of breads with less than 10% chia flour [85,90]. These differences may be due to the wheat flours used, the formulations used or the time of fermentation. In fact, the fermentation time was optimised in these two studies compared to the previous ones, which used a fixed fermentation time. This would indicate that the doughs enriched with chia flour would not support an excess of fermentation. Thus, the drop in volume and the consequent increase in the hardness of the breads may be due to the loss of strength of the dough and the breakage of the gluten network during fermentation, as shown by Steffolani et al. [11]. Other research, focused on other oilseeds such as sesame [37] or pumpkin [35] at levels above 14%, showed a decrease in the volume of the breads when they were enriched with flours of these seeds or protein concentrates made from them. However, Koca and Anil [86] did not observe differences in the volume of breads enriched with flaxseed flour at levels between 5% and 20%, which the authors attributed to the functionality of flaxseed gum. Likewise, Skrbic and Filipcev [38] did not observe differences either between the specific volumes or between the hardness of the breads, both white and whole, made with sunflower seeds. The differences with other investigations may be due to the incorporation of seeds and not flours, so the interaction of the components of the seeds with the components and the gluten is lower. However, a certain relaxation of the doughs was observed in this study since a decrease in the height/width bread ratio of the breads with 12% addition was observed. In general, the addition of oilseeds seems to have a smaller effect on the volume of the breads than the addition of oilseed flours. The incorporation of oilseeds as flour can have a positive effect or not modify the volume of the breads if they are incorporated in small proportions, while it shows a negative influence on the volume when the quantities exceed a certain level. This was exactly what Mentes et al. [98] observed in a study on the enrichment of wheat breads with ground flaxseed, where the negative effects began to be noticed after adding 20% flaxseed, and an increase of the volume with levels of 10% flaxseed. Similarly, Sayed-Ahmad et al. [99] observed that the incorporation of chia seed powder (full fat) at levels between 4–6% did not modify the hardness of the whole breads, while its inclusion in cakes (defatted venues) even reduced it.

Volume is one of the most reliable qualities of bread. Bread loaves with high volume often attract consumers. To improve the volume of oilseed-enriched breads, the addition of vital gluten [97] or the prehydration of the oilseeds has been proposed. When oilseeds are prehydrated, their mucilage interacts with the rest of the ingredients of the dough as a hydrocolloid would to reinforce the dough [11]. For the same reason, the use of additives (emulsifiers such as diacetyl tartaric acid ester of mono- and diglycerides (DATEM), or oxidants such as ascorbic acid) to strengthen the doughs could be tested. Zettel and Hitzmann [93] managed not to modify the volume of the breads and reduce their hardness when they used chia flour to substitute fats by increasing the moisture content of the formulation when the flour was incorporated.

The results also seem contradictory for gluten-free breads, something usual in this type of research since they depend on the flour/starch used as a base, the substitute for gluten and the hydration. It is known that the incorporation of insoluble fibers [100] and proteins [101] usually reduces bread volumes. In the case of oil, it can have a beneficial effect by reducing the consistency of the batters, which usually increases the volume of the bread to a certain limit [102]. All these investigations have been carried out on gluten-free breads with hydroxypropyl methylcellulose (HPMC), which can reach volumes similar to wheat breads and are very sensitive to changes in the formulation. In general, research shows a decrease in the volume of the breads and an increase in the hardness when chia flour was incorporated [41,94], although Constantini et al. [40] did not observe significant differences when they incorporated this flour to 10% in bread with buckwheat flour. However, in this study, no gluten substitute was used and the specific volume obtained was less than 1.5 mL/g, so that the doughs barely increased during the fermentation and baking processes. In contrast, the investigation of Sandri et al. [41] showed specific volumes of 1.7 mL/g with xanthan gum and carboxymethylcellulose, and specific volumes of 6 mL/g using HPMC were reached in the research of Steffolani et al. [94], which showed a greater reduction in volume with the incorporation of chia flour. Thus, changes are more appreciated when the specific volume of the control bread is greater.

The shelf life of enriched breads is another important aspect to consider. Although it has barely been studied in the research carried out until now, it is known that loaves with lower volumes and higher hardness generally harden more quickly [103]. Thus, the reduction of the volume of the loaves due to the incorporation of these seeds can accelerate the staling of the bread. Nevertheless, it is also known that lipids reduce the retrogradation phenomena of amylopectin [104] and, therefore, the rate of hardening of the loaves [105], so when oilseed flours are incorporated and the specific volume of the loaves is maintained, it is possible that the hardening of the loaves is delayed. It has been shown that the incorporation of chia flour can delay the retrogradation of amylopectin, responsible for the staling of the loaves [85]. The incorporation of ground flaxseed also retarded bread staling [98]. However, in long-life breads the oxidation of these lipids and their effect on the oxidation of other lipids present in the product must be considered, especially when they are incorporated as flour. Thus, breads enriched with flaxseed meal presented a higher level of peroxides after 8 weeks of storage [106]. An accelerated lipid oxidation, particularly polymerization compounds, was shown in chia-enriched biscuits [107]. These problems can be partially solved with the incorporation of antioxidant substances [108]. However, the possible combinations should also be considered since sesame seed extract has been shown to be effective in protecting oils from oxidation [109], and sesame oil has also been shown to be effective in delaying rancidity in oil of sunflower [110].

6. Sensory Properties

Although it has been proved that oilseeds can provide important health benefits, breads enriched with these seeds can only become commercially viable if consumers value their organoleptic properties. On this matter, multiple studies have evaluated the consumer acceptance of oilseed-enriched breads. In general, the oilseed enrichment of breads with and without gluten does not reduce the quality and acceptability to customers. Despite the addition of some seeds resulting in changes in the texture, colour and odour of the bread, there were no significant differences for crumb colour, crumb texture, quality and overall acceptability between the control and breads prepared with chia [11], sesame [37], pumpkin [36] and flax seeds [86].

As for breads enriched with oilseed flours, Verdú et al. [111] reported that bread enriched with 5% chia flour did not show significant sensory differences in respect to the control bread. This may be due to the low percentage of chia flour used. Coelho and Salas-Mellado [39] observed that the acceptability of enriched bread with oilseed flour is better than that of bread enriched with seeds, which they attributed to the higher specific volume of the bread with oilseed flour. However, in this research, they compared breads with 7.8 g/100 g of chia flour versus breads with 11 g/100 g of chia seeds, so the amount of chia could influence the evaluation of the consumers. Regardless, the use of prehydrated oilseed flours could be an option to improve the sensory properties of breads [11], probably due to the higher moisture level and less sensation of dryness, by releasing the mucilage prior to baking and increasing water retention during baking. In this respect, more research would be necessary in order to know the level of enrichment at which changes in the sensory properties become significant, since no agreement has been reached. The influence of enriched gluten-free breads with chia has been studied by authors such as Steffolani et al. [94] and Sandri et al. [41]. In both studies, no significant differences in terms of overall acceptability were reported in formulations with up to 15 g/100 g of chia flour or seeds.

In general, it can be affirmed that although the organoleptic characteristics of the breads change with the inclusion of oilseed seeds or their flours, there is no difference in the acceptability to consumers of these breads up to levels of 10–15%.

7. Conclusions

The incorporation of oilseeds improves the nutritional profile of bread, increasing its protein, fiber, vitamins, minerals, essential fatty acids and bioactive compounds. The use of oilseed mucilages has also been used successfully to replace the fat to produce healthier good quality breads. Therefore, the inclusion of these compounds in bakery products is of great interest, both in wheat products and in gluten-free products. However, it must be borne in mind that this inclusion modifies the rheology of the dough, depending on the way in which it is made (flour or seeds, prehydration or not) and the percentage used. Based on this, problems can result in changes to the final characteristics of the bread. To ensure the commercial success of these inclusions, it is necessary to consider the acceptability to consumers, which may vary depending on the type of inclusion and its percentage. In general, however, it seems that seed levels up to 15% can be obtained with good acceptability.

Author Contributions: B.D.L. and M.G. were responsible for the bibliographical search and writing of the article. All of the authors read and approved the final manuscript.

Acknowledgments: The authors acknowledge the financial support of the Spanish Ministry of Economy and Competitiveness (Project AGL2014-52928-C2) and the European Regional Development Fund (FEDER).

Conflicts of Interest: The authors declare no conflict of interest.

References

1. Muñoz, L.A.; Cobos, A.; Diaz, O.; Aguilera, J.M. Chia seeds: Microstructure, mucilage extraction and hydration. *J. Food Eng.* **2012**, *108*, 216–224. [CrossRef]
2. Kajla, P.; Sharma, A.; Sood, D.R. Flaxseed—A potential functional food source. *J. Food Sci. Technol.* **2015**, *52*, 1857–1871. [CrossRef] [PubMed]
3. USDA. *National Nutrient Database for Standard Reference*; USDA: Washington, DC, USA, 2018.
4. Anjum, F.M.; Nadeem, M.; Khan, M.I.; Hussain, S. Nutritional and therapeutic potential of sunflower seeds: A review. *Brit. Food J.* **2012**, *114*, 544–552. [CrossRef]
5. Goyal, A.; Sharma, V.; Upadhyay, N.; Gill, S.; Sihag, M. Flax and flaxseed oil: An ancient medicine & modern functional food. *J. Food Sci. Technol.* **2014**, *51*, 1633–1653. [PubMed]
6. Ullah, R.; Nadeem, M.; Khalique, A.; Imran, M.; Mehmood, S.; Javid, A.; Hussain, J. Nutritional and therapeutic perspectives of Chia (*Salvia hispanica* L.): A review. *J. Food Sci. Technol.* **2016**, *53*, 1750–1758. [CrossRef] [PubMed]
7. Seymen, M.; Uslu, N.; Türkmen, Ö.; Juhaimi, FA.; Özcan, M.M. Chemical compositions and mineral contents of some hull-less pumpkin seed and oils. *J. Am. Oil Chem. Soc.* **2016**, *93*, 1095–1099. [CrossRef]
8. Stevenson, D.G.; Eller, F.J.; Wang, L.; Jane, J.; Wang, T.; Inglett, G.E. Oil and tocopherol content and composition of pumpkin seed oil in 12 cultivars. *J. Agr. Food Chem.* **2007**, *55*, 4005–4013. [CrossRef] [PubMed]

9. Kuijsten, A.; Arts, I.C.W.; Van't Veer, P.; Hollman, P.C.H. The relative bioavailability of enterolignans in humans is enhanced by milling and crushing of flaxseed. *J. Nutr.* **2005**, *135*, 2812–2816. [CrossRef] [PubMed]
10. Edel, A.L.; Aliani, M.; Pierce, G.N. Stability of bioactives in flaxseed and flaxseed-fortified foods. *Food Res. Int.* **2015**, *77*, 140–155. [CrossRef]
11. Steffolani, E.; Martinez, M.M.; León, A.E.; Gómez, M. Effect of pre-hydration of chia (*Salvia hispanica* L.), seeds and flour on the quality of wheat flour breads. *LWT Food Sci. Technol.* **2015**, *61*, 401–406. [CrossRef]
12. Montesano, D.; Blasi, F.; Simonetti, M.S.; Santini, A.; Cossignani, L. Chemical and nutritional characterization of seed oil from *Cucurbita maxima* L. (var. Berrettina) pumpkin. *Foods* **2018**, *7*, 30. [CrossRef] [PubMed]
13. Marineli, R.D.; Moraes, E.A.; Lenquiste, S.A.; Godoy, A.T.; Eberlin, M.N.; Marostica, M.R. Chemical characterization and antioxidant potential of Chilean chia seeds and oil (*Salvia hispanica* L.). *LWT Food Sci. Technol.* **2014**, *59*, 1304–1310. [CrossRef]
14. Reyes-Caudillo, E.; Tecante, A.; Valdivia-Lopez, M.A. Dietary fibre content and antioxidant activity of phenolic compounds present in Mexican chia (*Salvia hispanica* L.) seeds. *Food Chem.* **2008**, *107*, 656–663. [CrossRef]
15. Vázquez-Obando, A.; Rosado-Rubio, G.; Chel-Guerrero, L.; Betancur-Ancona, D. Physicochemical properties of a fibrous fraction from chia (*Salvia hispanica* L.). *LWT Food Sci. Technol.* **2009**, *42*, 168–173.
16. Amato, M.; Caruso, M.C.; Guzzo, F.; Galgano, F.; Commisso, M.; Bochicchio, R.; Labella, R.; Favati, F. Nutritional quality of seeds and leaf metabolites of Chia (*Salvia hispanica* L.) from Southern Italy. *Eur. Food Res. Technol.* **2015**, *241*, 615–625. [CrossRef]
17. De Falco, B.; Amato, M.; Lanzotti, V. Chia seeds products: An overview. *Phytochem. Rev.* **2017**, *16*, 745–760. [CrossRef]
18. Rajesha, J.; Murthy, K.N.C.; Kumar, M.K.; Madhusudhan, B.; Ravishankar, G.A. Antioxidant potentials of flaxseed by in vivo model. *J. Agr. Food Chem.* **2006**, *54*, 3794–3799. [CrossRef] [PubMed]
19. Gouveia, L.D.V.; Cardoso, C.A.; de Oliveira, G.M.M.; Rosa, G.; Moreira, A.S.B. Effects of the intake of sesame seeds (*Sesamum indicum* L.) and derivatives on oxidative stress: A systematic review. *J. Med. Food* **2016**, *19*, 337–345. [CrossRef] [PubMed]
20. Shahidi, F.; Liyana-Pathirana, C.M.; Wall, D.S. Antioxidant activity of white and black sesame seeds and their hull fractions. *Food Chem.* **2006**, *99*, 478–483. [CrossRef]
21. Ghisoni, S.; Chiodelli, G.; Rocchetti, G.; Kane, D.; Lucini, L. UHPLC-ESI-QTOF-MS screening of lignans and other phenolics in dry seeds for human consumption. *J. Funct. Foods* **2017**, *34*, 229–236. [CrossRef]
22. Pająk, P.; Socha, R.; Broniek, J.; Królikowska, K.; Fortuna, T. Antioxidant properties, phenolic and mineral composition of germinated chia, golden flax, evening primrose, phacelia and fenugreek. *Food Chem.* **2019**, *275*, 69–76. [CrossRef]
23. Veronese, N.; Solmi, M.; Caruso, M.G.; Giannelli, G.; Osella, A.R.; Evangelou, E.; Maggi, S.; Fontana, L.; Stubbs, B.; Tzoulaki, I. Dietary fiber and health outcomes: An umbrella review of systematic reviews and meta-analyses. *Am. J. Clin. Nutr.* **2018**, *107*, 436–444. [CrossRef] [PubMed]
24. Anderson, J.W.; Baird, P.; Davis, R.H.; Ferreri, S.; Knudtson, M.; Koraym, A.; Waters, V.; Williams, C.L. Health benefits of dietary fiber. *Nutr. Rev.* **2009**, *67*, 188–205. [CrossRef] [PubMed]
25. Yashodhara, B.M.; Umakanth, S.; Pappachan, J.M.; Bhat, S.K.; Kamath, R.; Choo, B.H. Omega-3 fatty acids: A comprehensive review of their role in health and disease. *Postgrad. Med. J.* **2009**, *85*, 84–90. [CrossRef] [PubMed]
26. Toure, A.; Xu, X.M. Flaxseed lignans: Source, biosynthesis, metabolism, antioxidant activity, bio-active components, and health benefits. *Compr. Rev. Food Sci. Food Saf.* **2010**, *9*, 261–269. [CrossRef]
27. Namiki, M. The chemistry and physiological functions of sesame. *Food Rev. Int.* **1995**, *11*, 281–329. [CrossRef]
28. Hyvarinen, H.K.; Pihlava, J.M.; Hiidenhovi, J.A.; Hietaniemi, V.; Korhonen, H.J.T.; Ryhanen, E.L. Effect of processing and storage on the stability of flaxseed lignan added to bakery products. *J. Agr. Food Chem.* **2006**, *54*, 48–53. [CrossRef] [PubMed]
29. Teoh, S.L.; Lai, N.M.; Vanichkulpitak, P.; Vuksan, V.; Ho, H.; Chaiyakunapruk, N. Clinical evidence on dietary supplementation with chia seed (*Salvia hispanica* L.): A systematic review and meta-analysis. *Nutr. Rev.* **2018**, *76*, 219–242. [CrossRef] [PubMed]
30. Muñoz, L.A.; Cobos, A.; Diaz, O.; Aguilera, J.M. Chia seeds (*Salvia hispanica*): An ancient grain and a new functional food. *Food Rev. Int.* **2013**, *29*, 394–408. [CrossRef]

31. Prasad, K. Flaxseed and cardiovascular health. *J. Cardiovasc. Pharmacol.* **2009**, *54*, 369–377. [CrossRef] [PubMed]
32. Kanu, P.J.; Zhu, K.R.; Kanu, J.B.; Zhou, H.M.; Qian, H.F.; Zhu, K.X. Biologically active components and nutraceuticals in sesame and related products: A review and prospect. *Trends Food Sci. Technol.* **2007**, *19*, 599–608. [CrossRef]
33. Namiki, M. Nutraceutical functions of sesame: A review. *Crit. Rev. Food Sci. Technol.* **2007**, *47*, 651–673. [CrossRef] [PubMed]
34. Austria, J.A.; Aliani, M.; Malcolmson, L.J.; Dibrov, E.; Blackwood, D.P.; Maddaford, T.G.; Guzman, R.; Pierce, G.N. Daily choices of functional foods supplemented with milled flaxseed by a patient population over one year. *J. Funct. Foods* **2016**, *26*, 772–780. [CrossRef]
35. Mansour, E.H.; Dworschák, E.; Pollhamer, Zs.; Gergely, Á.; Hóvári, J. Pumplin and canola seed proteins and bread quality. *Acta Alimentaria* **1999**, *28*, 59–70.
36. El-Soukkary, F.A.H. Evaluation of pumpkin seed products for bread fortification. *Plant Food Hum. Nutr.* **2001**, *56*, 365–384. [CrossRef]
37. El-Adawy, T.A. Effect of sesame seed protein supplementation on the nutritional, physical, chemical and sensory properties of wheat flour bread. *Plant Food Hum. Nutr.* **1995**, *48*, 311–326. [CrossRef]
38. Škrbić, B.; Filipčev, B. Nutritional and sensory evaluation of wheat breads supplemented with oleic-rich sunflower seed. *Food Chem.* **2008**, *108*, 119–129. [CrossRef]
39. Coelho, M.S.; Salas-Mellado, M.M. Effects of substituting chia (*Salvia hispanica* L.) flour or seeds for wheat flour on the quality of the bread. *LWT Food Sci. Technol.* **2015**, *60*, 729–736. [CrossRef]
40. Costantini, L.; Lukšič, L.; Molinari, R.; Kreft, I.; Bonafaccia, G.; Manzi, L.; Merendino, N. Development of gluten-free bread using tartary buckwheat and chia flour rich in flavonoids and omega-3 fatty acids as ingredients. *Food Chem.* **2014**, *165*, 232–240. [CrossRef] [PubMed]
41. Sandri, L.T.B.; Santos, F.G.; Fratelli, C.; Capriles, V.D. Development of gluten-free bread formulations containing whole chia flour with acceptable sensory properties. *Food Sci. Nutr.* **2017**, *5*, 1021–1028. [CrossRef] [PubMed]
42. Gambus, H.; Mikulec, A.; Gambus, F.; Pisulewski, P. Perspectives of linseed utilization in baking. *Pol. J. Food Nutr. Sci.* **2004**, *13*, 21–27.
43. Seczyk, L.; Swieca, M.; Dziki, D.; Anders, A.; Gawlik-Dziki, U. Antioxidant, nutritional and functional characteristics of wheat bread enriched with ground flaxseed hulls. *Food Chem.* **2017**, *214*, 32–38. [CrossRef] [PubMed]
44. Martin, J.; Geisel, T.; Maresch, C.; Krieger, K.; Stein, J. Inadequate nutrient intake in patients with celiac disease: Results from a German dietary survey. *Digestion* **2013**, *87*, 240–246. [CrossRef] [PubMed]
45. Saturni, L.; Ferretti, G.; Bacchetti, T. The gluten-free diet: Safety and nutritional quality. *Nutrients* **2010**, *2*, 16–34. [CrossRef] [PubMed]
46. Theethira, T.G.; Dennis, M. Celiac disease and the gluten-free diet: Consequences and recommendations for improvement. *Dig. Dis.* **2015**, *33*, 175–182. [CrossRef] [PubMed]
47. Vici, G.; Belli, L.; Biondi, M.; Polzonetti, V. Gluten free diet and nutrient deficiencies: A review. *Clin. Nutr.* **2016**, *35*, 1236–1241. [CrossRef] [PubMed]
48. Caruso, R.; Pallone, F.; Stasi, E.; Romeo, S.; Monteleone, G. Appropriate nutrient supplementation in celiac disease. *Ann. Med.* **2013**, *45*, 522–531. [CrossRef] [PubMed]
49. Kulai, T.; Rashid, M. Assessment of nutritional adequacy of packaged gluten-free food products. *Can. J. Diet. Pract. Res.* **2014**, *75*, 186–190. [CrossRef] [PubMed]
50. Mazzeo, T.; Cauzzi, S.; Brighenti, F.; Pellegrini, N. The development of a composition database of gluten-free products. *Public Health Nutr.* **2015**, *18*, 1353–1357. [CrossRef] [PubMed]
51. Miranda, J.; Lasa, A.; Bustamante, M.A.; Churruca, I.; Simon, E. Nutritional differences between a gluten-free diet and a diet containing equivalent products with gluten. *Plant Food Hum. Nutr.* **2014**, *69*, 182–187. [CrossRef] [PubMed]
52. Missbach, B.; Schwingshackl, L.; Billmann, A.; Mystek, A.; Hickelsberger, M.; Bauer, G.; Konig, J. Gluten-free food database: The nutritional quality and cost of packaged gluten-free foods. *Peer J.* **2015**, *3*, e1337. [CrossRef] [PubMed]

53. Regula, J.; Cerba, A.; Suliburska, J.; Tinkov, A.A. In vitro bioavailability of calcium, magnesium, iron, zinc, and copper from gluten-free breads supplemented with natural additives. *Biol. Trace Elem. Res.* **2017**, *182*, 140–146. [CrossRef] [PubMed]
54. Swieca, M.; Regula, J.; Suliburska, J.; Zlotek, U.; Gawlik-Dziki, U. Effects of gluten-free breads, with varying functional supplements, on the biochemical parameters and antioxidant status of rat serum. *Food Chem.* **2015**, *182*, 268–274. [CrossRef] [PubMed]
55. Hung, W.L.; Liao, C.D.; Lu, W.C.; Ho, C.T.; Hwang, L.S. Lignan glycosides from sesame meal exhibit higher oral bioavailability and antioxidant activity in rat after nano/submicrosizing. *J. Funct. Food* **2016**, *23*, 511–522. [CrossRef]
56. Nesbitt, P.D.; Lam, Y.; Thompson, L.U. Human metabolism of mammalian lignan precursors in raw and processed flaxseed. *Am. J. Clin. Nutr.* **1999**, *69*, 549–555. [CrossRef] [PubMed]
57. Malcolmson, L.J.; Przybylski, R.; Duan, J.K. Storage stability of milled flaxseed. *J. Am. Oil Chem. Soc.* **2000**, *77*, 235–238. [CrossRef]
58. Schorno, A.L.; Manthey, F.A.; Hall, C.A., III. Effect of particle size and sample size on lipid stability of milled flaxseed (*Linum usitatissimum* L.). *J. Food Process Preserv.* **2010**, *34*, 167–179. [CrossRef]
59. Fedeniuk, R.W.; Biliaderis, C.G. Composition and physicochemical properties of linseed (*Linum-usitatissimum* L.) mucilage. *J. Agr. Food Chem.* **1994**, *42*, 240–247. [CrossRef]
60. Korus, J.; Witczak, T.; Ziobro, R.; Juszczak, L. Flaxseed (*Linum usitatissimum* L.) mucilage as a novel structure forming agent in gluten-free bread. *LWT Food Sci. Technol.* **2015**, *62*, 257–264. [CrossRef]
61. Cui, W.; Mazza, G.; Biliaderis, C.G. Chemical-structure, molecular-size distributions, and rheological properties of flaxseed gum. *J. Agr. Food Chem.* **1994**, *42*, 1891–1895. [CrossRef]
62. Capitani, M.I.; Ixtaina, V.Y.; Nolasco, S.M.; Tomas, M.C. Microstructure, chemical composition and mucilage exudation of chia (*Salvia hispanica* L.) nutlets from Argentina. *J. Sci. Food Agr.* **2013**, *93*, 3856–3862. [CrossRef] [PubMed]
63. Capitani, M.I.; Corzo-Rios, L.J.; Chel-Guerrero, L.A.; Betancur-Ancona, D.A.; Nolasco, S.M.; Tomas, M.C. Rheological properties of aqueous dispersions of chia (*Salvia hispanica* L.) mucilage. *J. Food Eng.* **2015**, *149*, 70–77. [CrossRef]
64. Timilsena, Y.P.; Adhikari, R.; Kasapis, S.; Adhikari, B. Molecular and functional characteristics of purified gum from Australian chia seeds. *Carbohydr. Polym.* **2016**, *136*, 128–136. [CrossRef] [PubMed]
65. Capitani, M.I.; Nolasco, S.M.; Tomas, M.C. Stability of oil-in-water (O/W) emulsions with chia (*Salvia hispanica* L.) mucilage. *Food Hydrocolloids* **2016**, *61*, 537–546. [CrossRef]
66. Mazza, G.; Biliaderis, C.G. Functional-properties of flax seed mucilage. *J. Food Sci.* **1989**, *54*, 1302–1305. [CrossRef]
67. Campos, B.E.; Ruivo, T.D.; Scapim, M.R.D.; Madrona, G.S.; Bergamasco, R.D. Optimization of the mucilage extraction process from chia seeds and application in ice cream as a stabilizer and emulsifier. *LWT Food Sci. Technol.* **2016**, *65*, 874–883. [CrossRef]
68. Borneo, R.; Aguirre, A.; León, A.E. Chia (*Salvia hispanica* L.) gel can be used as egg or oil replacer in cake formulations. *J. Am. Diet. Assoc.* **2010**, *110*, 946–949. [CrossRef] [PubMed]
69. Felisberto, M.H.F.; Wahanik, A.L.; Gomes-Ruffi, C.R.; Clerici, M.T.P.S.; Chang, Y.K.; Steel, C.J. Use of chia (*Salvia hispanica* L.) mucilage gel to reduce fat in pound cakes. *LWT - Food Sci. Technol.* **2015**, *63*, 1049–1055. [CrossRef]
70. Fernandes, S.S.; Salas-Mellado, M.D.L.M. Addition of chia seed mucilage for reduction of fat content in bread and cakes. *Food Chem.* **2017**, *227*, 237–244. [CrossRef] [PubMed]
71. Fernandes, S.S.; Salas-Mellado, M.D.L.M. Development of mayonnaise with substitution of oil or egg yolk by the addition of chia (*Salvia hispanica* L.) mucilage. *J. Food Sci.* **2018**, *83*, 74–83. [CrossRef] [PubMed]
72. Rosell, C.M.; Rojas, J.A.; de Barber, C.B. Influence of hydrocolloids on dough rheology and bread quality. *Food Hydrocoll.* **2001**, *15*, 75–81. [CrossRef]
73. Ribotta, P.D.; Perez, G.T.; Leon, A.E.; Añon, M.C. Effect of emulsifier and guar gum on micro structural, rheological and baking performance of frozen bread dough. *Food Hydrocoll.* **2004**, *18*, 305–313. [CrossRef]
74. Guarda, A.; Rosell, C.M.; Benedito, C.; Galotto, M.J. Different hydrocolloids as bread improvers and antistaling agents. *Food Hydrocoll.* **2004**, *18*, 241–247. [CrossRef]
75. Mandala, I.; Karabela, D.; Kostaropoulos, A. Physical properties of breads containing hydrocolloids stored at low temperature. I. Effect of chilling. *Food Hydrocoll.* **2007**, *21*, 1397–1406. [CrossRef]

76. Anton, A.A.; Artfield, S.D. Hydrocolloids in gluten-free breads: A review. *Int. J. Food Sci. Nutr.* **2008**, *59*, 11–23. [CrossRef] [PubMed]
77. Sabanis, D.; Tzia, C. Effect of hydrocolloids on selected properties of gluten-free dough and bread. *Food Sci Technol. Int.* **2011**, *17*, 279–291. [CrossRef] [PubMed]
78. Brennan, C.S.; Blake, D.E.; Ellis, P.R.; Schofield, J.D. Effects of guar galactomannan on wheat bread microstructure and on the in vitro and in vivo digestibility of starch in bread. *J. Cereal Sci.* **1996**, *24*, 151–160. [CrossRef]
79. Lu, Z.X.; Walker, K.Z.; Muir, J.G.; O'Dea, K. Arabinoxylan fibre improves metabolic control in people with Type II diabetes. *Eur. J. Clin. Nutr.* **2004**, *58*, 621–628. [CrossRef] [PubMed]
80. Salgado-Cruz, M.D.; Ramirez-Miranda, M.; Diaz-Ramirez, M.; Alamilla-Beltran, L.; Calderon-Dominguez, G. Microstructural characterisation and glycemic index evaluation of pita bread enriched with chia mucilage. *Food Hydrocoll.* **2017**, *69*, 141–149. [CrossRef]
81. Sluimer, P. Principles of Breadmaking. In *Functionality of Raw Materials and Process Steps*; AACC: St Paul. MN, USA, 2005.
82. Liu, X.; Li, T.; Liu, B.Y.; Zhao, H.F.; Zhou, F.; Zhang, B.L. An external addition of soy protein isolate hydrolysate to sourdough as a new strategy to improve the quality of chinese steamed bread. *J. Food Qual.* **2016**, *39*, 3–12. [CrossRef]
83. Perez-Carrillo, E.; Chew-Guevara, A.A.; Heredia-Olea, E.; Chuck-Hernandez, C.; Serna-Saldivar, S.O. Evaluation of the functionality of five different soybean proteins in hot-press wheat flour tortillas. *Cereal Chem.* **2015**, *92*, 98–104. [CrossRef]
84. Gomez, M.; Ronda, F.; Blanco, C.A.; Caballero, P.A.; Apesteguia, A. Effect of dietary fibre on dough rheology and bread quality. *Eur. Food Res. Technol.* **2003**, *216*, 51–56. [CrossRef]
85. Iglesias-Puig, E.; Haros, M. Evaluation of performance of dough and bread incorporating chia (*Salvia hispanica* L.). *Eur. Food Res. Technol.* **2013**, *237*, 865–874. [CrossRef]
86. Koca, A.F.; Anil, M. Effect of flaxseed and wheat flour blends on dough rheology and bread quality. *J. Sci. Food Agr.* **2007**, *87*, 1172–1175.
87. Moreira, R.; Chenlo, F.; Torres, M.D. Effect of chia (*Salvia hispanica* L.) and hydrocolloids on the rheology of gluten-free doughs based on chestnut flour. *LWT Food Sci. Technol.* **2013**, *50*, 160–166. [CrossRef]
88. Matthews, R.H.; Sharpe, E.J.; Clark, W.M. The use of some oilseed flours in bread. *Cereal Chem.* **1970**, *47*, 181–188.
89. Svec, I.; Hruskova, M.; Jurinova, I. Pasting characteristics of wheat-chia blends. *J. Food Eng.* **2016**, *172*, 25–30. [CrossRef]
90. Verdú, S.; Vásquez, F.; Ivorra, E.; Sánchez, A.J.; Barat, J.M.; Grau, R. Physicochemical effects of chia (*Salvia hispanica*) seed flour on each wheat bread-making process phase and product storage. *J. Cereal Sci.* **2015**, *66*, 67–73. [CrossRef]
91. Agyare, K.K.; Addo, K.; Xiong, Y.L.; Akoh, C.C. Effect of structured lipid on alveograph characteristics, baking and textural qualities of soft wheat flour. *J. Cereal Sci.* **2005**, *42*, 309–316. [CrossRef]
92. Ribotta, P.D.; Arnulphi, S.A.; Leon, A.E.; Añon, M.C. Effect of soybean addition on the rheological properties and breadmaking quality of wheat flour. *J. Sci. Food Agr.* **2005**, *85*, 1889–1896. [CrossRef]
93. Zettel, V.; Hitzmann, B. Chia (*Salvia hispanica* L.) as fat replacer in sweet pan breads. *Int. J. Food Sci. Technol.* **2016**, *51*, 1425–1432. [CrossRef]
94. Steffolani, E.; de la Hera, E.; Perez, G.; Gomez, M. Effect of chia (*Salvia hispanica* L.) addition on the quality of gluten-free bread. *J. Food Qual.* **2014**, *5*, 309–317. [CrossRef]
95. Zhou, J.M.; Liu, J.F.; Tang, X.Z. Effects of whey and soy protein addition on bread rheological property of wheat flour. *J. Texture Stud.* **2018**, *49*, 38–46. [CrossRef] [PubMed]
96. Pareyt, B.; Finnie, S.M.; Putseys, J.A.; Delcour, J.A. Lipids in bread making: Sources, interactions, and impact on bread quality. *J. Cereal Sci.* **2011**, *54*, 266–279. [CrossRef]
97. Luna Pizarro, P.; Almeida, E.L.; Coelho, A.S.; Sammán, N.C.; Hubinger, M.D.; Chang, Y.K. Functional bread with n-3 alpha linolenic acid from whole chia (*Salvia hispanica* L.) flour. *J. Food Sci. Technol.* **2015**, *52*, 4475–4482. [CrossRef] [PubMed]
98. Mentes, Ö.; Bakkalbasi, E.; Ercan, R. Effect of the use of ground flaxseed on quality and chemical composition of bread. *Food Sci. Technol. Int.* **2008**, *14*, 299–306. [CrossRef]

99. Sayed-Ahmad, B.; Talou, T.; Straumite, E.; Sabovics, M.; Kruma, Z.; Saad, Z.; Hijazi, A.; Merah, O. Evaluation of nutritional and technological attributes of whole wheat based bread fortified with chia flour. *Foods* **2018**, *7*, 135. [CrossRef] [PubMed]
100. Martinez, M.M.; Diaz, A.; Gomez, M. Effect of different microstructural features of soluble and insoluble fibres on gluten-free dough rheology and bread-making. *J. Food Eng.* **2014**, *142*, 49–56. [CrossRef]
101. Sahagún, M.; Gómez, M. Assessing influence of protein source on characteristics of gluten-free breads optimising their hydration level. *Food Bioprocess Technol.* **2018**, *11*, 1686–1694. [CrossRef]
102. Mancebo, C.M.; Martinez, M.M.; Merino, C.; de la Hera, E.; Gomez, M. Effect of oil and shortening in rice bread quality: Relationship between dough rheology and quality characteristics. *J. Texture Stud.* **2017**, *48*, 597–606. [CrossRef] [PubMed]
103. Gómez, M.; Oliete, B.; Pando, V.; Ronda, F.; Caballero, P.A. Effect of fermentation conditions on bread staling kinetics. *Eur. Food Res. Technol.* **2008**, *226*, 1379–1387. [CrossRef]
104. Eliasson, A.C.; Ljunger, G. Interactions between amylopectin and lipid additives during retrogradation in a model system. *J. Sci. Food Agr.* **1988**, *44*, 353–361. [CrossRef]
105. Rogers, D.E.; Zeleznak, K.J.; Lai, C.S.; Hoseney, R.C. Effect of native lipids, shortening, and bread moisture on bread firming. *Cereal Chem.* **1988**, *65*, 398–401.
106. Conforti, F.D.; Davis, S.F. The effect of soya flour and flaxseed as a partial replacement for bread flour in yeast bread. *Int. J. Food Sci. Technol.* **2006**, *41*, 95–101. [CrossRef]
107. Mesias, M.; Holgado, F.; Marquez-Ruiz, G.; Morales, F.J. Risk/benefit considerations of a new formulation of wheat-based biscuit supplemented with different amounts of chia flour. *LWT Food Sci. Technol.* **2016**, *73*, 528–535. [CrossRef]
108. Villanueva, E.; Rodriguez, G.; Aguirre, E.; Castro, V. Influence of antioxidants on oxidative stability of the oil Chia (*Salvia hispanica* L.) by rancimat. *Sci. Agropecu.* **2017**, *8*, 19–27. [CrossRef]
109. Konsoula, Z.; Liakopoulou-Kyriakides, M. Effect of endogenous antioxidants of sesame seeds and sesame oil to the thermal stability of edible vegetable oils. *LWT Food Sci. Technol.* **2010**, *43*, 1379–1386. [CrossRef]
110. Ghosh, M.; Upadhyay, R.; Mahato, D.K.; Mishra, H.N. Kinetics of lipid oxidation in omega fatty acids rich blends of sunflower and sesame oils using Rancimat. *Food Chem.* **2019**, *272*, 471–477. [CrossRef] [PubMed]
111. Verdú, S.; Barat, J.M.; Grau, R. Improving bread-making processing phases of fibre-rich formulas using chia (*Salvia hispanica*) seed flour. *LWT Food Sci. Technol.* **2017**, *84*, 419–425. [CrossRef]

© 2018 by the authors. Licensee MDPI, Basel, Switzerland. This article is an open access article distributed under the terms and conditions of the Creative Commons Attribution (CC BY) license (http://creativecommons.org/licenses/by/4.0/).

Article

Consumer-Driven Improvement of Maize Bread Formulations with Legume Fortification

Luís M. Cunha [1], Susana C. Fonseca [1], Rui C. Lima [2], José Loureiro [3], Alexandra S. Pinto [4], M. Carlota Vaz Patto [5] and Carla Brites [4,*]

1. GreenUPorto & LAQV/REQUIMTE, DGAOT, Faculdade de Ciências da Universidade do Porto, Campus de Vairão, Rua da Agrária 747, 4485-646 Vila do Conde, Portugal
2. Sense Test, Lda., R. Zeferino Costa 341, 4400-345 Vila Nova de Gaia, Portugal
3. Patrimvs Indústria, SA, Zona Industrial da Abrunheira, 33/34 Apt 184, Vila Chã, 6270-187 Guarda, Portugal
4. INIAV-Instituto Nacional de Investigação Agrária e Veterinária, 2780-157 Oeiras, Portugal
5. ITQB NOVA-Instituto de Tecnologia Química e Biológica António Xavier, Universidade Nova de Lisboa, Av República, Apartado 127, 2781-901 Oeiras, Portugal
* Correspondence: carla.brites@iniav.pt; Tel.: +351-214-403-500

Received: 24 May 2019; Accepted: 25 June 2019; Published: 29 June 2019

Abstract: The fortification of maize bread with legume flour was explored in order to increase the protein content of the traditional Portuguese bread '*broa*', comprised of more than 50% maize flour. The optimization of legume incorporation (pea, chickpea, faba bean, lentil), considering the influence of different maize flours (traditional-white, traditional-yellow, hybrid-white, hybrid-yellow), on consumer liking and sensory profiling of '*broa*' was studied. A panel of 60 naïve tasters evaluated twenty different breads, divided in four sets for each legume flour fortification, each set including four breads with varying maize flour and a control (no legume). Tasters evaluated overall liking and the sensory profile through a check-all-that-apply ballot. Crude protein and water content were also analyzed. There were no significant differences in overall liking between the different types of legumes and maize. The incorporation of chickpea flour yields a sensory profile that most closely resembles the control. The protein content increased, on average, 21% in '*broa*', with legume flours having the highest value obtained with faba bean incorporation (29% increase). Thus, incorporation of legume flours appears to be an interesting strategy to increase bread protein content, with no significant impact on consumer liking and the '*broa*' bread sensory profile.

Keywords: maize bread; legume fortification; pea; chickpea; faba bean; lentil; protein content

1. Introduction

Bread, a cereal-based product, is an important part of the human diet but rich in easily digested carbohydrates that are associated with a high glycemic index food consumption, which is a health concern for present consumers. Wheat is the most frequently used cereal for bread making in many parts of the world, but other common ingredients of bread are maize and rye [1]. An interesting alternative to wheat bread, with a lower glycemic index, is a maize-based bread named '*broa*' [2].

'*Broa*' is a Portuguese bread comprised of more than 50% maize flour, mixed with either wheat and/or rye flours, and highly consumed in the northern and central regions of Portugal [3]. Several types of '*broa*' can be produced, depending on the type of maize variety and blending of the flours used, with regional maize landraces (normally open pollinated varieties, OPV) being considered more suitable than hybrid varieties for bread production [4,5]. Maize flour parameters related to the consumer perceived quality of Portuguese '*broa*' bread, based on eleven regional OPV maize landraces, were evaluated [6]. The study revealed similar hedonic assessments (appearance, odor, texture, flavor, color, global appreciation, and cohesiveness) of '*broa*' bread among specialty landraces of maize flours and the lowest scores for 'broa' bread from commercial (hybrid variety) maize flour. In that study, commercial flour presented the highest mean diameter and a larger flour particle distribution range of all the tested maize varieties.

The traditional bread making process used to prepare 'broa' consists of mixing maize flour (sieved whole meal flour), wheat and/or rye flour, hot water, yeast, salt, and leavened dough from a previous bread (acting as the sourdough) [3]. After mixing and resting, the dough is baked in a wood-fired oven. This empirical process leads to an ethnic product highly appreciated for its distinctive sensory characteristics (unique flavor and texture) and provides an interesting source of nutritional value [2,3,7]. The microbiological profile of flours used to manufacture 'broa' bread and the microbial phenomena of dough fermentation and storage for 'broa' bread was studied in [7–9].

A gluten-free 'broa' bread, with modification of the traditional composite maize/rye wheat flour, was tested and considered satisfactory for its sensory quality and bread making technology ability [10]. Maize dough has no gluten proteins, which enables it to hold the gas produced during fermentation in a viscoelastic network, leading to a compact bread with crumb-like texture and low specific volume [5].

Legumes are generally rich in protein and fiber and low in fat, and are considered to have a high nutritional value and key role in preventing metabolic diseases, such as diabetes mellitus and coronary heart diseases [11,12]. Thus, legumes can contribute significantly to the protein fortification of cereal-based products to align them with the high vegetal protein diet trend [13]. Legume-enriched bread may be amenable with claims such as 'source of', 'high' or 'increased of' vegetal protein according with Reg EC 1924/2006 [14]. Despite nutritional enrichment, the organoleptic quality of legume-fortified cereal foods was significantly different and tends to decline [15–17] when compared with formulations based exclusively on cereals. Thus, the assessment of consumer acceptability is essential to promote the incorporation of legume flours in cereal-based products. The pre-treatment of grain legumes (roasting, cooking, or fermentation) influences the composition and protein properties of grain legumes and, consequently, the characteristics of dough and bread fortified with legume flours [18].

The objectives of this work were to select the optimal maize bread formulation with legume fortification (part of maize flour replaced by legume flour) accessed through overall consumer liking and a check-all-that-apply (CATA) profiling evaluation, in order to obtain bread claiming to have a high protein content that is also nutritiously enriched and well accepted by consumers.

2. Materials and Methods

2.1. Experimental Procedure

The base formulation of the 'broa' bread included 700 g·kg^{-1} maize, 200 g·kg^{-1} rye, 100 g·kg^{-1} wheat flour, 28 g·kg^{-1} sugar (wt/wt flour basis), 17.6 g·kg^{-1} salt (wt/wt flour basis), 10 g·kg^{-1} dry yeast (wt/wt flour basis), 100 g·kg^{-1} sourdough (wt/wt flour basis), and 100% (vol/wt flour basis) water, as described previously by Brites et al. [2].

Twenty breads were produced following a 4-block design combining four legume flours: chickpea—CH (*Cicer arietinum*), faba bean—FB (*Vicia faba*), lentil—LC (*Lens culinaris*), pea—PS (*Pisum sativum*), and control—C (without legume flour) with four different maize flours: hybrid white—IW, hybrid yellow—IY, regional white—RW, regional yellow—RY. The regional whole maize flours were obtained after milling the grain in an artisanal water mill with millstones (Moinhos do Inferno, Viseu), and the hybrid flours correspond to commercial maize flour (Nacional type 175). Each of the four blocks corresponds to the replacement of 10% of the maize flour by one of the legume flours (CH, FB, LC, PS) and a control (C) sample with no replacement.

The 'broa' bread making process, performed at Patrimvs Indústria, a bakery industry in Portugal, was previously described in [2] and consisted of mixing the maize flour with 80% of the boiling salted water and kneading for 5 min (Ferneto AEF035). The dough was allowed to rest and cool to 27 °C, and the remaining ingredients (sugar, salt, dry yeast, sourdough), including 20% of the water, were added. The dough was again kneaded for 8 min and left to rest for bulk fermentation at 25 °C for 90 min. After fermentation, the dough was manually molded into 400 g balls and baked in an oven (Matador, Werner & Pfleiderer Lebensmitteltechnik GmbH, Dinkelsbühl, Germany) at 270 °C for 40 min (Figure 1a).

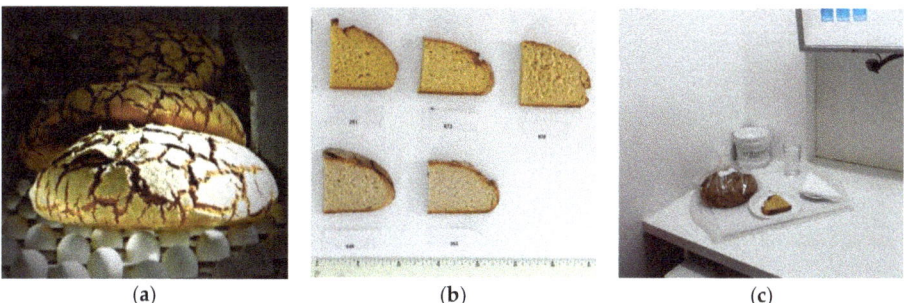

(a) (b) (c)

Figure 1. (a) *'Broa'* bread after baking; (b) Slices of *'broa'* bread samples for sensory evaluation; (c) Individual booth with sample for sensory evaluation.

Maize breads produced by Patrimvs Indústria were packed and dispatched the day before each of the sensory evaluation sessions, according to the different legume flour blocks.

2.2. Sensory Analysis

Sixty naïve tasters who consumed bread regularly were recruited for a descriptive profiling test from Sense Test's (an independent Sensory Analysis Laboratory in Portugal) consumer database. A sociodemographic characterization of consumers was performed. The company ensures data protection and confidentiality through the authorization 2063/2009 awarded by the National Data Protection Commission and an accomplished internal code of conduct.

Sensory evaluation was carried out at Sense Test in a special room equipped with individual booths in accordance with ISO standard 8589:2007 [19], with personnel and panel leader following ISO standards 13300-1:2006 [20] and 13300-2:2006 [21].

Four sessions were set up and, for each session, five different samples of maize bread were produced: (i) white regional maize with legume flour; (ii) white hybrid maize with legume flour; (iii) yellow regional maize with legume flour; (iv) yellow hybrid maize with legume flour; and a control sample made with yellow hybrid maize with no legume flour (Figure 1b). *'Broa'* breads were cut halfway and 2 cm-thick slices were cut from the central portion. A serving of one slice (≈100 g) was presented to each taster on white disposable plastic plates, identified by a three-digit random number, at the individual booths under normal white lighting (Figure 1c). Panelists were provided with a porcelain spittoon, a glass of bottled natural water, and unsalted crackers. All panelists were instructed to chew a piece of cracker and to rinse their mouth with water before testing each sample. Panelists were free to swallow or spit both samples and crackers.

Within a session, each participant received all the samples following a monadic sequential (one at a time) presentation, with a balanced sample serving order to compensate for possible carryover effects [22]. Overall liking was evaluated using the classical 9-point hedonic scale, going from 1—"dislike extremely" to 9—"like extremely" [23]. For each sample, overall liking scoring was immediately followed by the evaluation of the sensory profile, through a check-all-that-apply (CATA) methodology to reduce bias [24,25].

Participants were invited to profile each sample over a CATA ballot, structured according to 6 sensory dimensions (Whole bread appearance (WBA), Slice appearance (SA), Texture at touch (TT), Mouth texture (MT), Aroma (A), and Flavor (F)). Table 1 presents the total 51 attributes, according to the respective sensory dimension. The CATA ballot was generated after discussion between the authors based on previous research [26]. This list was presented in two different orders to the panelists, following a direct and an inverse alphabetic order within each dimension [25]. The CATA ballot,

"Please select out of the following list of terms those that characterize the tasted sample", was answered as a "yes/no" response scale, indicating if they recognized the presence or absence of such attributes. This option increased the focus of respondents on each attribute [27].

Table 1. Dimensions and descriptive attributes used in check-all-that-apply (CATA) ballot.

Dimension	Number of Attributes	Descriptive Attributes
Whole bread appearance (WBA)	3	Dark toasted crust, Uniform crust, Strongly cracked crust
Slice appearance (SA)	6	Whitish crumb, Yellowish crumb, High porous crumb, Large grain crumb, Small grain crumb, Homogeneous porosity crumb
Texture at touch (TT)	5	Moist, Soft, Granular crumb, Compact, Sticky
Mouth texture (MT)	11	Crumbly, Hard crust, Crunchy crust, Elastic crust, Dry crumb, Sticky, Moist, Elastic, Granular, Soft, Lumps together
Aroma (A)	11	Intense, Acid, Maize, Rye, Pea, Faba bean, Chickpea, Lentil, Yeast, Burnt, Moldy
Flavour (F)	15	Toasted crust, Intense, Sweet, Acid, Bitter, Salty, Bland, Pea, Faba bean, Chickpea, Lentil, Maize, Rye, Moldy, Yeast

2.3. Protein and Water Content Determination

The bread samples were prepared according to the AACC 62-05.01 method [28], water content by the ISO 712:2009 [29], and protein content by the combustion method of ISO 16634-2/TS:2016 [30], calculated by multiplying nitrogen concentration by a conversion factor of 5.7.

2.4. Statistical Analysis

Data analyses were performed using the XL-STAT 2019® system software (Addinsoft, New York, NY, USA). To synthesize the results of the overall liking test, descriptive statistics with mean and standard error (SE) for each session, corresponding to the different legume flour breads and the control sample, were computed. A two-way (type of maize and type of legume) ANOVA with blocks (tasters), and the Fisher-LSD test for multiple comparison (differences in the liking of each of the enriched breads), was applied at a 95% confidence level. A two-way (type of maize and type of legume) ANOVA, and Fisher-LSD test for multiple comparison, was applied to evaluate differences between both the protein and water content of the different 'broa' breads (with interaction as the error) at a 95% confidence level.

For CATA evaluations, the Cochran test was applied to identify which descriptive attributes were discriminating among samples [31]. Subsequently, the frequency of use of each attribute was determined, calculating the number of panelists who have used each attribute to describe the samples. Over this frequency matrix, a correspondence analysis (CA) was applied. Such analysis provides a sensory map of the bread samples, allowing the perception of the similarities and differences between samples and their sensory characteristics [32–34]. Multidimensional alignment (MDA) was also performed to determine the correlation between the descriptive attributes and the bread samples in the full-dimensional space of the CA, providing complete information about the relationship between products and attributes [35].

3. Results and Discussion

Generally, the overall liking for all types of breads presented high values of acceptability, around 7—"like very much" (Table 2). Breads with legume incorporation tended to have a somewhat lower acceptability than the control samples (Table 2). Significant differences in the overall liking of the control samples used in each session were identified. This was probably due to industrial variability, as batches were produced in different time periods. To overcome this effect, differences of overall liking of each of the legume-incorporated breads, and the respective control bread, were calculated. From the two-way ANOVA model with panelists as blocks, no significant effects were found for the type of legume, type of maize as well, as for the interaction of both factors ($p > 0.05$). Despite no significant effect, the bread incorporating chickpea flour with hybrid white maize was the only one presenting an average overall liking (6.97 ± 0.16) above the corresponding control sample (6.77 ± 0.17) (Table 2).

Table 2. Mean values ± standard error (SE) of overall liking of control breads (C-IY) and breads combining four legume flours (CH—chickpea, FB—faba bean, LC—lentil, and PS—pea) with 4 maize flours (IW—hybrid white, IY—hybrid yellow; RW—regional white, RY—regional yellow), using a 9-point scale, going from 1—"dislike extremely" to 9—"like extremely".

Session	Sample	n	Overall Liking	
1	C-IY	60		6.77 (±0.17) [a]
	CH-IW	60	6.97 (±0.16) [a]	
	CH-IY	60	6.53 (±0.18) [a]	6.63 (±0.09) (n = 240)
	CH-RW	60	6.50 (±0.20) [a]	
	CH-RY	60	6.52 (±0.16) [a]	
2	C-IY	60		7.22 (±0.15) [a]
	FB-IW	60	6.95 (±0.12) [a]	
	FB-IY	60	7.20 (±0.14) [a]	6.93 (±0.08) (n = 240)
	FB-RW	60	6.90 (±0.16) [a]	
	FB-RY	60	6.68 (±0.17) [a]	
3	C-IY	60		7.22 (±0.18) [a]
	LC-IW	60	6.98 (±0.17) [a]	
	LC-IY	60	7.00 (±0.14) [a]	6.95 (±0.08) (n = 240)
	LC-RW	60	6.97 (±0.17) [a]	
	LC-RY	60	6.83 (±0.17) [a]	
4	C-IY	60		7.08 (±0.15) [a]
	PS-IW	60	6.88 (±0.15) [a]	
	PS-IY	60	6.92 (±0.14) [a]	6.89 (±0.07) (n = 240)
	PS-RW	60	7.00 (±0.15) [a]	
	PS-RY	60	6.75 (±0.16) [a]	

[a] within each session, the same letter indicates no significant differences between 'broas', according to the Fisher-LSD test ($p > 0.05$).

Results from the CATA profiling by Cochran test yielded both discriminating and non-discriminating attributes. The non-discriminating attributes identified across all blocks are presented in Table 3. It is important to highlight that the non-discriminating terms associated with the legume incorporation, such as the chickpea, lentil, and pea aroma and the chickpea and faba bean flavor, indicated that the presence of the respective legume flours was not noticeable to consumers.

Table 3. Dimensions and descriptive attributes of non-discriminating attributes across all blocks.

Dimension	Descriptive Attributes
Slice appearance (SA)	Homogeneous porosity crumb
Mouth texture (MT)	Crunchy crust, Elastic crust, Hard crust
Aroma (A)	Chickpea, Intense, Lentil, Moldy, Pea, Yeast
Flavor (F)	Chickpea, Faba bean, Salty, Intense

Figure 2 shows the configurations of the samples and discriminating descriptive terms in the first and second dimensions of the CA analysis, applied to the CATA data of bread samples. This configuration explained 78.06% of the total variance of the experimental data. From the analysis of Figure 2, it is possible to observe that the samples were grouped according to flour maize type, and that the hybrid white (IW) and regional white (RW) maize were strongly associated with (SA) Whitish crumb, in contrast to hybrid yellow (IY) and regional yellow (RY) maize breads that were intensely associated with (SA) Yellowish crumb, as expected. Data yielded a very high consistency on the profiling of the control samples across all 4 blocks (circle lined in Figure 2), correlated with (SA) Yellowish crumb, (MT) Dry crumb, (TT) Compact, (MT) Crumbly, (A) Maize, and (F) Maize. The incorporation of chickpea flour yielded the sensory profile that most closely resembled the control samples. The regional maize flours were closer associated with sticky and moist descriptors, related with bread crumb cohesiveness (sticky texture at touch and in mouth), and apparent humidity (perceived moisture at touch and in mouth).

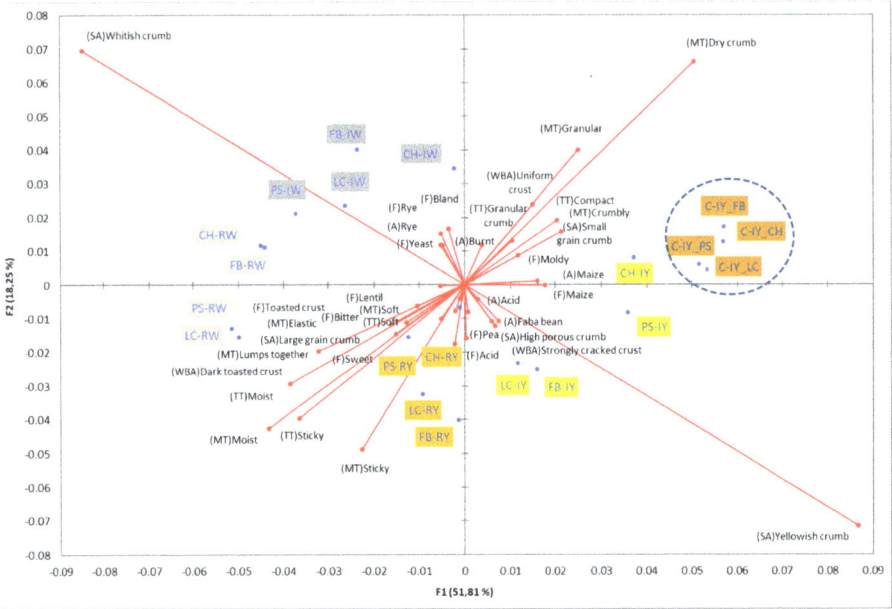

Figure 2. Sensory profiling for bread fortification with legume flours, produced combining 4 legume flours (CH—chickpea, FB—faba bean, LC—lentil, and PS—pea) and a C—control (without legume flour) with 4 maize flours (IW—hybrid white, IY—hybrid yellow, RW—regional white, RY—regional yellow).

Table 4 shows the results from the cosines of the angles of 'broa' bread samples, with the significant descriptive attributes, resulting from MDA analysis. Angles below 45° indicate a significant positive correlation between the projection of the sample and the projection of the attribute into the CA space, while angles above 135° indicate a significant negative correlation between the projection of the sample and the projection of the attribute [35]. From this analysis, it was possible to depict, in a more detailed way, the differences in bread samples and their relationship with the descriptive terms, since MDA is a statistical procedure that takes into account all the dimensions. The association between (SA) Whitish crumb and (SA) Yellowish crumb, according to the white and yellow maize respectively, was again evident in Table 4. The positive association with the acid flavor for faba bean, lentil, and pea was highlighted in this analysis.

Table 4. Multidimensional alignment (MDA) for bread fortification with legume flours, produced combining 4 legume flours (CH—chickpea, FB—faba bean, LC—lentil, and PS—pea) with 4 maize flours (IW—hybrid white, IY—hybrid yellow, RW—regional white, and RY—regional yellow). Significant correlations between samples and attributes are depicted with the bold bars.

Table 4. Cont.

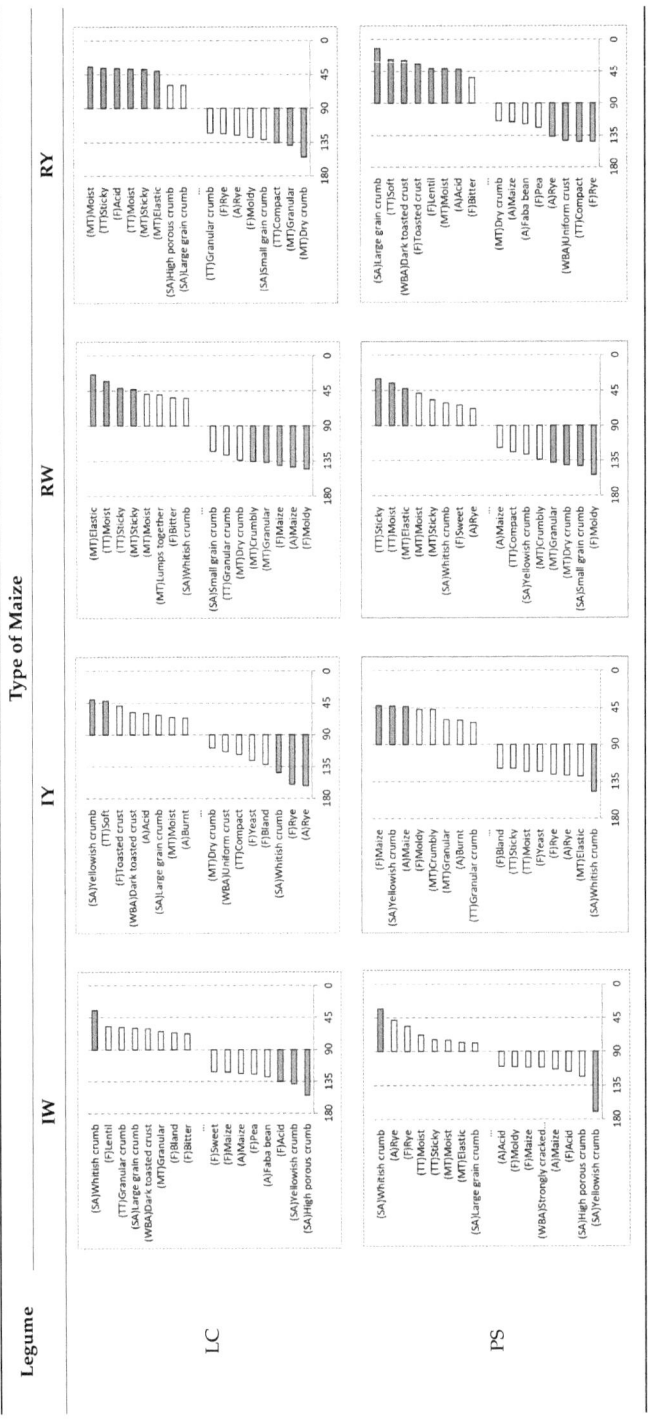

Table 5 presents the protein and water content of the 'broa' bread produced, combining the 4 legume flours and the control (without legume flour) with the 4 maize flours. Legume fortification significantly increased the 'broa' bread protein content, as expected. The mean protein content of the bread without legume incorporation was 56.2 (±1.3) g·kg^{-1} and increased, on average, by 21% in 'broa' with the incorporation of legume flours. The highest increase in protein content was obtained with faba bean (29%) incorporation rising to 72.3 (±2.4) g·kg^{-1}, compared to the control without legume incorporation. In terms of protein content per dry weight, the differences were even higher between legume fortification and the control bread. The faba bean incorporation increased the protein content to 32% (118.3 (±1.8) g·kg^{-1} dry basis), followed by lentil incorporation of 22% (108.8 (±0.8) g·kg^{-1} dry basis), and chickpea and pea incorporation (17% and 16%, respectively). The lowest water content was obtained from the bread with pea incorporation (362.5 (±5.6) g·kg^{-1}) and the highest from faba bean incorporation (389.5 (±12.4) g·kg^{-1}). A significantly lower water content was obtained from the hybrid varieties of bread in comparison with the regional varieties of bread.

Table 5. Mean values ± SE of protein and water content of the 'broa' breads produced combining 4 legume flours (CH—chickpea, FB—faba bean, LC—lentil, and PS—pea) and C—control (without legume flour) with 4 maize flours (IW—hybrid white, IY—hybrid yellow, RW—regional white, RY—regional yellow).

'Broa' Sample	Protein Content (g·kg^{-1})	Protein Content in a Dry Basis (g·kg^{-1} Dry Basis)	Water Content (g·kg^{-1})
C	56.2 (±1.3) [a]	89.5 (±1.0) [a]	372.0 (±10.7) [a,b]
CH	65.6 (±1.7) [b]	105.0 (±1.4) [b]	375.3 (±10.5) [a,b]
FB	72.3 (±2.4) [c]	118.3 (±1.8) [d]	389.5 (±12.4) [b]
LC	67.7 (±1.4) [b]	108.8 (±0.8) [c]	377.3 (±12.1) [a,b]
PS	66.3 (±0.4) [b]	104.0 (±1.1) [b]	362.5 (±5.6) [a]
IW	66.1 (±1.9) [a,b]	104.0 (±4.1) [a]	363.4 (±11.4) [a]
IY	68.0 (±3.7) [b]	106.0 (±5.3) [a,b]	358.8 (±3.6) [a]
RW	65.3 (±2.4) [a,b]	107.0 (±4.5) [b]	388.8 (±6.8) [b]
RY	63.0 (±3.0) [a]	103.4 (±4.9) [a]	390.2 (±4.0) [b]

[a,b,c,d] Similar letters indicate homogeneous groups according to the Fisher-LSD test ($p > 0.05$).

4. Conclusions

Incorporation of legume flours appears to be an interesting strategy to increase bread protein content without decreasing consumer liking. However, further research should also consider the impact of legume incorporation on the glycemic index of maize bread. Regarding the sensory profile method, one can observe that the CATA was an appropriate method to describe the maize bread formulations with legume fortification. The breads were produced in the bakery chain of Patrimvs S.A, with the intended positive effect of studying a real-life market production situation, but the industrial scale imposed limitations, such as the restricted choice of commercially available legumes flours. Major changes in the 'broa' sensory profile appear related to apparent humidity (perceived moisture at touch and in mouth) and bread crumb cohesiveness (sticky texture at touch and in mouth). Incorporation of chickpea flour lead to liking scores closer to the control formulation. The incorporation of chickpea flour yielded the sensory profile that most closely resembled the control. Faba flour incorporation lead to 'broa' breads with the highest protein content.

These results can be seen as an opportunity for the bakery industry to develop new products that will respond to the growing consumer demand for high-protein food.

Author Contributions: Conceptualization, L.M.C., C.V.P, C.B.; methodology, L.M.C., R.C.L., C.B.; formal analysis, L.M.C., S.C.F.; resources, J.L.; writing—original draft preparation, L.M.C., S.C.F., C.B.; writing—review and editing, L.M.C., S.C.F., R.C.L., J.L., A.S.P, C.V.P., C.B.; supervision, L.M.C., C.B.; project administration, C.V.P., C.B.; funding acquisition, C.V.P., C.B.

Funding: This research was funded by EU's Seventh Framework Programme for Research, Technological Development and Demonstration under the grant agreement n. 613551, LEGATO project.

Acknowledgments: Authors acknowledge PATRIMVS Indústria SA, Portugal by industrially producing bread using different maize and legume species and Sense Test Lda, Portugal for recruitment of the sensory evaluation panel". Authors acknowledge José B. Cunha, from the Oporto British School, for the revision of English usage and grammar.

Conflicts of Interest: The authors declare no conflict of interest.

References

1. Ohimain, E.I. Recent advances in the production of partially substituted wheat and wheatless bread. *Eur. Food Res. Technol.* **2015**, *240*, 257–271. [CrossRef]
2. Brites, C.M.; Trigo, M.J.; Carrapiço, B.; Alviña, M.; Bessa, R.J. Maize and resistant starch enriched breads reduce postprandial glycemic responses in rats. *Nutr. Res.* **2011**, *31*, 302–308. [CrossRef] [PubMed]
3. Brites, C.; Haros, M.; Trigo, M.J.; Islas, R.P. Maíz. In *De Tales Harinas, Tales Panes: Granos, Harinas y Productos de Panificación n Iberoamérica*; León, A.E., Rosell, C.M., Báez, H., Eds.; ISEKI-Food: Vienna, Austria, 2007; pp. 74–121. ISBN 9789871311071.
4. Vaz Patto, M.C.; Moreira, P.M.; Carvalho, V.; Pego, S. Collecting maize (*Zea mays* L. convar. *mays*) with potential technological ability for bread making in Portugal. *Genet. Res. Crop Evol.* **2007**, *54*, 1555–1563. [CrossRef]
5. Vaz Patto, M.C.; Alves, M.L.; Almeida, N.F.; Santos, C.; Moreira, P.M.; Satovic, Z.; Brites, C. Is the bread making technological ability of Portuguese traditional maize landraces associated with their genetic diversity? *Maydica* **2009**, *54*, 297–311. [CrossRef]
6. Carbas, B.; Vaz Patto, M.C.; Bronze, M.R.; Bento-da-Silva, A.; Trigo, M.J.; Brites, C. Maize flour parameters that are related to the consumer perceived quality of 'broa' specialty bread. *Food Sci. Technol. Campinas* **2016**, *36*, 259–267. [CrossRef]
7. Rocha, J.M.; Malcata, F.X. Behavior of the complex micro-ecology in maize and rye flour and mother-dough for *broa* throughout storage. *J. Food Qual.* **2016**, *39*, 218–233. [CrossRef]
8. Rocha, J.M.; Malcata, F.X. Microbiological profile of maize and rye flours, and sourdough used for the manufacture of traditional Portuguese bread. *Food Microbiol.* **2012**, *31*, 72–88. [CrossRef]
9. Rocha, J.M.; Malcata, F.X. Microbial ecology dynamics in portuguese *broa* sourdough. *J. Food Qual.* **2016**, *39*, 634–648. [CrossRef]
10. Brites, C.; Trigo, M.J.; Santos, C.; Collar, C.; Rosell, C.M. Maize-based gluten-free bread: Influence of processing parameters on sensory and instrumental quality. *Food Bioproc. Technol.* **2010**, *3*, 707–715. [CrossRef]
11. Tharanathan, R.N.; Mahadevamma, S. Grain legumes—A boon to human nutrition. *Trends Food Sci. Technol.* **2003**, *14*, 507–518. [CrossRef]
12. Boye, J.; Zare, F.; Pletch, A. Pulse proteins: Processing, characterization, functional properties and applications in food and feed. *Food Res. Int.* **2010**, *43*, 414–431. [CrossRef]
13. Pinto, A.; Guerra, M.; Carbas, B.; Pathania, S.; Castanho, A.; Brites, C. Challenges and opportunities for food processing to promote consumption of pulses. *Rev. Ciências Agrárias* **2016**, *39*, 571–582. [CrossRef]
14. The European Parliament and the Council of the European Union. Regulation (EC) No 1924/2006 of the European Parliament and of the Council of 20 December 2006 on nutrition and health claims made on foods. *Off. J. Eur. Union* **2006**, 1–17.
15. Fenn, D.; Lukow, O.M.; Humphreys, G.; Fields, P.G.; Boye, J.I. Wheat-legume composite flour quality. *Int. J. Food Prop.* **2010**, *13*, 381–393. [CrossRef]
16. Anyango, J.O.; de Kock, H.L.; Taylor, J.R.N. Evaluation of the functional quality of cowpea-fortified traditional African sorghum foods using instrumental and descriptive sensory analysis. *LWT-Food Sci. Technol.* **2011**, *44*, 2126–2133. [CrossRef]
17. Du, S.K.; Jiang, H.; Yu, X.; Jane, J. Physicochemical and functional properties of whole legume flour. *LWT-Food Sci. Technol.* **2014**, *55*, 308–313. [CrossRef]
18. Baik, B.K.; Han, I.H. Cooking, Roasting, and Fermentation of Chickpeas, Lentils, Peas, and Soybeans for Fortification of Leavened Bread. *Cereal Chem.* **2012**, *89*, 269–275. [CrossRef]

19. International Organization for Standardization. *ISO Standard 8589:2007 Sensory Analysis—General Guidance for the Design of Test Rooms*; International Organization for Standardization: Geneva, Switzerland, 2007.
20. International Organization for Standardization. *ISO Standards 13300-1:2006 Sensory Analysis—General Guidance for the Staff of a Sensory Evaluation Laboratory—Part 1: Staff Responsibilities*; International Organization for Standardization: Geneva, Switzerland, 2006.
21. International Organization for Standardization. *ISO Standards 13300-2:2006 Sensory Analysis—General Guidance for the Staff of a Sensory Evaluation Laboratory—Part 2: Recruitment and Training of Panel Leaders*; International Organization for Standardization: Geneva, Switzerland, 2006.
22. Macfie, H.J.; Bratchell, N.; Greenhoff, K.; Vallis, L.V. Designs to balance the effect of order of presentation and first-order carry-over effects in hall tests. *J. Sens. Stud.* **1989**, *4*, 129–148. [CrossRef]
23. Peryam, D.R.; Pilgrim, F.J. Hedonic scale method of measuring food preferences. *Food Technol.* **1957**, *11*, 9–14.
24. Ares, G.; Jaeger, S.R. Check-all-that-apply questions: Influence of attribute order on sensory product characterization. *Food Qual. Prefer.* **2013**, *28*, 141–153. [CrossRef]
25. King, S.C.; Meiselman, H.L.; Carr, B.T. Measuring emotions associated with foods: Important elements of questionnaire and test design. *Food Qual. Prefer.* **2013**, *28*, 8–16. [CrossRef]
26. Seabra Pinto, A.; Brites, C.; Vaz Patto, C.; Cunha, L. Do Consumers' Value the New Use of Legumes? An Experimental Auction with Legume Fortified Maize Bread. In *Book of Abstracts, Proceedings of the Second International Legume Society Conference Legumes for a Sustainable World, Troia, Portugal, 11–14 October 2016*; New University of Lisbon: Lisabon, Portugal; p. 54.
27. Jaeger, S.R.; Cadena, R.S.; Torres-Moreno, M.; Antunez, L.; Vidal, L.; Gimenez, A.; Ares, G. Comparison of check-all-that-apply and forced-choice yes/no question formats for sensory characterisation. *Food Qual. Prefer.* **2014**, *35*, 32–40. [CrossRef]
28. American Association of Cereal Chemists. *AACC Method 62-05.01 Preparation of Sample: Bread*; American Association of Cereal Chemists: Eagan, MN, USA, 2002.
29. International Organization for Standardization. *ISO 712:2009. Cereals and Cereal Products—Determination of Moisture Content—Reference Method*; International Organization for Standardization: Geneva, Switzerland, 2009.
30. International Organization for Standardization. *ISO 16634/TS:2016. Food Products—Determination of the Total Nitrogen Content by Combustion According to the Dumas Principle and Calculation of the Crude Protein Content*; International Organization for Standardization: Geneva, Switzerland, 2016.
31. Parente, M.E.; Manzoni, A.V.; Ares, G. External preference mapping of commercial antiaging creams based on consumers' responses to a check-all-that-apply question. *J. Sens. Stud.* **2011**, *26*, 158–166. [CrossRef]
32. Ares, G.; Barreiro, C.; Deliza, R.; Giménez, A.; Gàmbaro, A. Application of a check-all-that-apply question to the development of chocolate milk desserts. *J. Sens. Stud.* **2010**, *25*, 67–86. [CrossRef]
33. Ares, G.; Deliza, R.; Barreiro, C.; Giménez, A.; Gámbaro, A. Comparison of two sensory profiling techniques based on consumer perception. *Food Qual. Prefer.* **2010**, *21*, 417–426. [CrossRef]
34. Ares, G.; Varela, P.; Rado, G.; Giménez, A. Identifying ideal products using three different consumer profiling methodologies. Comparison with external preference mapping. *Food Qual. Prefer.* **2011**, *22*, 581–591. [CrossRef]
35. Meyners, M.; Castura, J.C.; Carr, B.T. Existing and new approaches for the analysis of CATA data. *Food Qual. Prefer.* **2013**, *30*, 309–319. [CrossRef]

© 2019 by the authors. Licensee MDPI, Basel, Switzerland. This article is an open access article distributed under the terms and conditions of the Creative Commons Attribution (CC BY) license (http://creativecommons.org/licenses/by/4.0/).

Article

The Use of Upcycled Defatted Sunflower Seed Flour as a Functional Ingredient in Biscuits

Simona Grasso [1,*], Ese Omoarukhe [1], Xiaokang Wen [2], Konstantinos Papoutsis [2] and Lisa Methven [3]

1. School of Agriculture, Policy and Development, University of Reading, Reading RG6 6AR, UK
2. School of Agriculture and Food Science, University College Dublin, D04 V1W8 Dublin, Ireland
3. Department of Food and Nutritional Sciences, University of Reading, Reading RG6 6AP, UK
* Correspondence: simona.grasso@ucdconnect.ie

Received: 28 June 2019; Accepted: 27 July 2019; Published: 1 August 2019

Abstract: Defatted sunflower seed flour (DSSF) is an upcycled by-product of sunflower oil extraction, rich in protein, fibre and antioxidants. This study assessed the instrumental and sensory quality of biscuits enriched with DSSF at 18% and 36% *w/w* as a replacement for wheat flour. Measurements included colour, texture, total phenolic content (TPC) and antioxidant capacity. Sensory analysis was carried out with Quantitative Descriptive Analysis (QDA). The inclusion of DSSF significantly increased the protein content of the biscuits, as well as the TPC and antioxidant capacity of the biscuits. The resulting products were significantly darker, less red and less yellow with increasing DSSF levels, while hardness (measured instrumentally) increased. Sensory results agreed with colour measurements, concluding that DSSF biscuits were more "Brown" than the control, and with texture measurements where biscuits with 36% DSSF had a significantly firmer bite. In addition, DSSF biscuits at 36% inclusion had higher QDA scores for "Off-note" and the lowest scores for "Crumbly" and "Crumb aeration". DSSF biscuits at 18% inclusion were similar to the control in most parameters and should be considered for further developments. These results show the potential of the upcycled DSSF by-product as a novel, sustainable and healthy food ingredient.

Keywords: biscuits; upcycled food by-products; defatted sunflower seed flour; sensory QDA; TPA; colour; antioxidant capacity; protein enrichment; functional foods; valorisation

---

1. Introduction

Sunflower (*Helianthus annuus* L.) is one of the three most cultivated oil crops in the world [1]. The main by-product of the oil extraction process, which can constitute up to 36% of the mass of the processed seeds [1], is the so-called sunflower meal or cake. This by-product has a high protein content (40–50%) [2] and is used primarily in ruminant feed [1].

The sunflower cake contains essential amino acids (such as lysine, methionine, cystine, tryptophan), minerals, B group vitamins [1] and has a high antioxidant potential [3], making this product interesting as human food. On the other hand, some limitations include a high insoluble fibre content, the residue solvents used for oil extraction in the cake [3] and the presence of anti-nutrients such as protease inhibitors, saponins and arginase inhibitor [4].

Steam explosion, involving high pressure and high temperature, has recently been used on various substrates and by-products to break insoluble fibre into smaller soluble dietary fibre units [5,6], to decompose some anti-nutrients [7] and as a sterilisation method [8]. The US company Planetarians uses steam explosion on sunflower cake to produce a commercially available food grade defatted sunflower seed flour (DSSF) without the need for purification steps [9]. Within the context of circular bio-economy, there is a growing interest in the food industry to use inexpensive upcycled by-products to partially replace flour, fat or sugar in bakery products to achieve value-added, nutritionally-enriched and sustainable foods [10].

The aim of the present work is to use DSSF in biscuits, substituting it for 18% and 36% of wheat flour, and investigating the effects that the DSSF inclusion might have on the quality of the biscuits, both from an instrumental and sensory point of view.

2. Materials and Methods

2.1. Materials

Wheat flour (composition from manufacturer: fat 1.7%, carbohydrate 74%, fibre 3.8%, protein 9.9%), sugar, sunflower oil, cocoa powder, sodium bicarbonate and sodium chloride used for biscuit formulations were supermarket own-label from a local retailer. The DSSF (composition from manufacturer: fat 1%, carbohydrate 48%, fibre 18%, protein 35%), obtained after steam explosion and milling, was donated by the company Planetarians (Palo Alto, CA, USA).

2.2. Biscuit Preparation

Short dough biscuits were prepared according to the modified method of Kuchtová, Karovičová, Kohajdová, Minarovičová and Kimličková [10]. Control biscuits were manufactured without DSSF, while DSSF biscuits were made substituting respectively 18% and 36% of wheat flour for DSSF. In the control recipe, the oil (26.5 g) and sugar (35 g) were mixed at medium speed for 5 min using a mixer (Major Titanium KM020, Kenwood, London, UK). Water (62 g) was added and mixed for 30 s at low speed. Then the remaining ingredients (flour 90 g, cocoa 10 g, sodium bicarbonate 1.1 g and salt 0.9 g) were added and mixed for 2 min at low speed. In the 18% and 36% DSSF recipes, respectively 18% of flour (16.4 g) or 36% of flour (32.8 g) was replaced with DSSF. The dough was sheeted to a 4 mm thickness with a Rondo table model dough sheeter (Rondo, Burgdorf, Switzerland) and cut by hand with a 55 mm diameter round cutter. The biscuits were baked on aluminium trays in a ventilated oven (Kwick_Co, Salva, Gipuzkoa, Spain) for 15 min at 190 °C. After 30 min of cooling time, the biscuits were vacuum packed and stored until further analysis.

2.3. Proximate Analysis

The moisture, protein, fat and ash content of the wheat flour, DSSF and biscuits were determined using methods from the Association of Official Analytical Chemists (AOAC) [11]. The total content of carbohydrates was calculated by difference: 100 − (moisture + ash + protein + fat).

2.4. Determination of Total Phenolic Content and Antioxidant Capacity

The extracts for the determination of total phenolic content (TPC) and antioxidant capacity were prepared according to Ajila et al. [12] with some modifications. Briefly, 1 g of sample was mixed with 20 mL of absolute methanol and left at ambient temperature for 1 h. Subsequently, the mixture was centrifuged (1.5× g, 10 min) and the supernatant was collected and used for the determination of TPC and antioxidant capacity.

2.4.1. Total Phenolic Content (TPC)

TPC was measured for the DSSF, wheat flour and biscuits according to Singleton and Rossi [13] with some modifications. Briefly, 2.5 mL of 10% (v/v) Folin–Ciocalteu reagent was mixed with 0.5 mL of sample. After 3 min of incubation, 2 mL of 7.5% (w/v) Na_2CO_3 was added to the mixture and incubated in the dark at room temperature for 1 h. The absorbance of the solution was measured at 765 nm. The results were expressed as mg of gallic acid equivalents per g of sample dry weight (mg GAE/g).

2.4.2. Antioxidant Capacity

The 2,2-Diphenyl-1-picrylhydrazyl (DPPH) radical scavenging capacity was determined according to Vamanu and Nita [14] and Papoutsis et al. [15]. Briefly, 2850 mL of DPPH solution was mixed with 150 µL of extract. The mixture was left to stand for 30 min in the dark. The absorbance was measured at 515 nm. Results were expressed mg Trolox equivalents per g (mg TE/g).

The cupric reducing antioxidant capacity (CUPRAC) was determined according to Apak et al. [16] with some modifications. Briefly, 1 mL of 10 mM copper chloride (II) was mixed with 1 mL of 7.5 mM neocuproine solution and 1 mL of NH_4Ac buffer (pH 7.0). Subsequently, 1.1 mL of sample was added to this mixture. The mixture was incubated at room temperature for 1.5 h before measuring the absorbance at 450 nm. Results were expressed as mg Trolox equivalents per g (mg TE/g).

2.5. Physical Analyses

The width and thickness of at least 10 biscuits per batch and per recipe were measured with a digital calliper after baking. The spread ratio of the biscuits was calculated dividing width by thickness [17].

Hardness was measured using a texture analyser TA-XT2i (Stable Micro Systems, London, UK) equipped with a three-point bending rig (HDP/3PB). Hardness was the maximum resistance of each biscuit against a rounded edge blade and occurred when the sample began to break.

Water holding capacity (WHC) and oil-adsorption capacity (OAC) was measured for wheat flour and DSSF. WHC was determined according to Sudha et al. [18] with slight modifications. Aliquots of 0.05 g of DSSF or wheat flour were mixed with 1 mL water in a microcentrifuge tube, centrifuged at 13,000× g for 30 min, and the excess water was decanted. The sample was weighed, and WHC was expressed as g water/g dry weight. OAC was similarly determined, by using sunflower oil instead of water. OAC was expressed as g oil/g dry weight.

2.6. Colour

The colour of the wheat flour, DSSF and biscuits was measured using a colorimeter (CR-400, Konica, Minolta, Japan), calibrated using a white standard plate. The values measured were L* (white 100/black 0), a* values (red positive/green negative) and b* values (yellow positive/blue negative). Colour was measured for 10 biscuits in each batch. The total colour difference (ΔE) was calculated according to the equation:

$$\Delta E = [(a^* - a_0^*)^2 + (b^* - b_0^*)^2 + (L^* - L_0^*)^2]^{1/2}$$

2.7. Sensory Evaluation

Sensory profiling of biscuits was conducted by a panel of nine trained panellists (eight female, one male, mean age 47 years). A consensus vocabulary of 29 descriptors was developed to characterise the samples; under the modalities appearance (5), aroma (4), taste and flavour (8), mouthfeel (5) and after-effect descriptors (7), using reference standards where required. The purpose of the sensory profiling was to provide a consistent measure for changes in biscuit descriptors occurring with change in formulation. Descriptor scoring was done using unstructured line-scales (scale 0–100) using the Compusense® software (Compusense, ON, Canada). Panellists were seated in individual testing booths under artificial daylight. Samples (one biscuit per person per sample) were presented in a balanced order, randomly allocated and single-blinded using three-digit number codes. Panellists were asked to taste at least half of the portion size. Warm filtered water was used as a palate cleanser and the time delay between samples (post after-effects scoring) was 30 s. Biscuit scoring was carried out in duplicates on two consecutive days.

2.8. Statistical Analysis

For instrumental measurements, the experiment was repeated three times on three different days. Statistical analyses were carried out using analysis of variance (ANOVA) or independent t-tests with the software SPSS (V24, SPSS Inc., Chicago, IL, USA). The determination of significance among the control, 18% DSSF, and 36% DSSF biscuits was conducted by Tukey's post hoc multiple comparison test at a significance level of $p < 0.05$. For sensory data, a two-way ANOVA was used. The panellists were fitted as random effects and the samples were fixed effects. The treatment effects (samples and assessors) were tested against the panellist by assessor interaction.

3. Results and Discussion

3.1. Physical and Chemical Properties of Defatted Sunflower Seed Flour

The physical and chemical properties of wheat flour and DSSF are reported in Table 1. The DSSF ingredient presented a lower moisture and a fourfold higher protein content compared to wheat flour. Fat and ash content were also significantly higher compared to wheat flour.

In terms of colour, DSSF was significantly darker than wheat flour, with higher redness and blueness values. The hydration properties of DSSF fell in the same range reported by other authors on apple pomace [18,19]. DSSF presented a threefold higher WHC than wheat flour, possibly due to the high content of soluble dietary fibre. Ash content is a good indicator of mineral content and according to the literature [20], sunflower oil cake on a dry basis contains 0.48% calcium, 0.84% phosphorus, 0.44% magnesium and 3.49% potassium.

Table 1. Physical and chemical properties of defatted sunflower seed flour and wheat flour.

Parameters		Wheat Flour	DSSF
Moisture (%)		10.22 ± 0.02 [a]	4.59 ± 0.02 [b]
Protein (%)		9.8 ± 0.02 [b]	38.01 ± 0.01 [a]
Fat (%)		1.62 ± 0.01 [b]	1.84 ± 0.03 [a]
Ash (%)		0.94 ± 0.05 [b]	7.19 ± 0.03 [a]
WHC (g water/g dry weight)		0.69 ± 0.14 [b]	2.21 ± 0.18 [a]
OAC (g oil/g dry weight)		0.87 ± 0.03 [b]	1.25 ± 0.06 [a]
Colour	L*	93.93 ± 0.36 [a]	62.99 ± 0.12 [b]
	a*	−0.79 ± 0.06 [b]	1.47 ± 0.02 [a]
	b*	11.45 ± 0.15 [b]	14.38 ± 0.07 [a]

Data are expressed as means ±SD of duplicate or triplicate assays. Values with the same letter in the same row are not significantly different at $p < 0.05$. DSSF, defatted sunflower seed flour; OAC, oil-adsorption capacity; WHC, water holding capacity.

Results for TPC, DPPH and CUPRAC are shown in Figure 1. TPC in DSSF was 16.54 mg GAE/g, while TPC for wheat flour was significantly lower at 5.59 mg GAE/g. These values are higher than those reported on apple pomace (1.1 mg GAE/g dry weight) [19] and beetroot pomace (up to 3.8 mg GAE/g dry weight) [21]. Similarly to TPC results, DSSF had higher antioxidant capacity measured by DPPH and CUPRAC assays compared to wheat flour. Previous studies have shown that sunflower flour is a good source of phenolic compounds including chlorogenic, caffeic, p-hydroxybenzoic, p-coumaric, cinamic, m-hydroxybenzoic, vanillic, syringic, transcinnamic, isoferulic and sinapic acids, which are compounds with high antioxidant capacity [22]. On the other hand, wheat flour has been reported to have very low polyphenol content [23], which justifies its lower antioxidant capacity compared to DSSF.

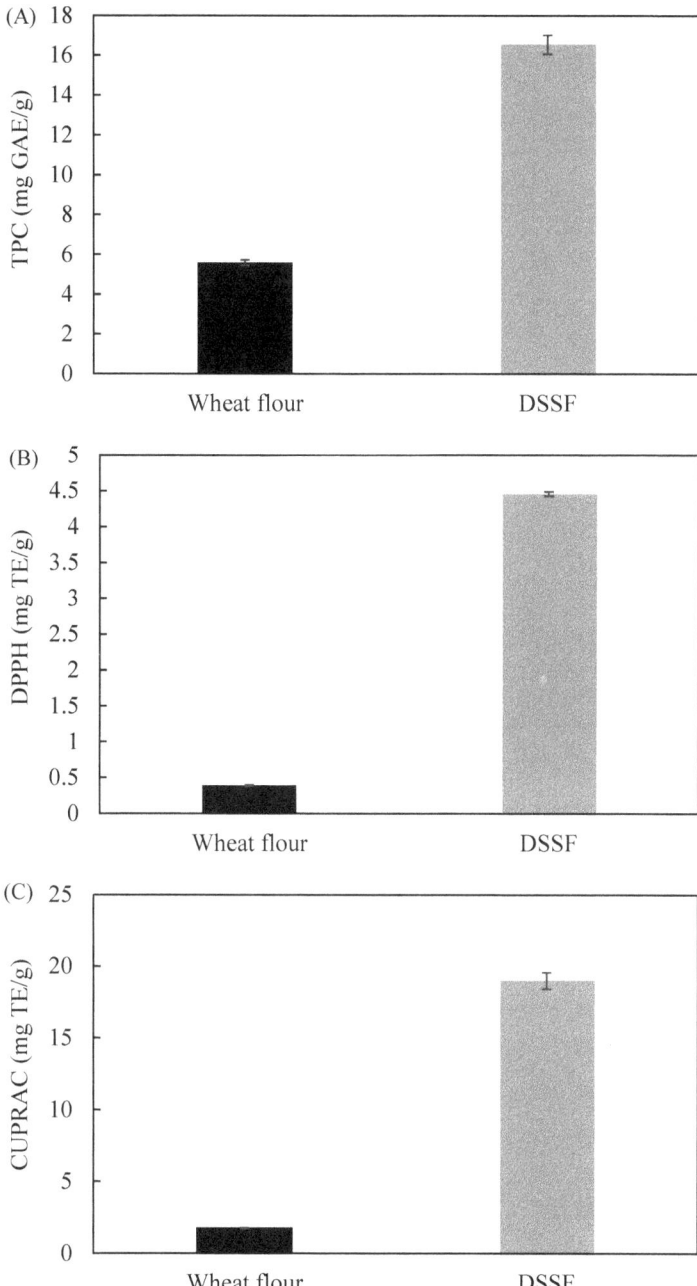

Figure 1. Total phenolic content (TPC) (**A**) 2,2-Diphenyl-1-picrylhydrazyl (DPPH) (**B**) and cupric reducing antioxidant capacity (CUPRAC) (**C**) values of wheat flour and defatted sunflower seed flour (DSSF). Data are expressed as means ± SD (n = 3).

3.2. Effect of Defatted Sunflower Seed Flour on the Physical and Chemical Properties of Biscuits

The physical properties of biscuits made replacing wheat flour with 18% and 36% of DSSF are presented in Table 2. When compared to control biscuits, the diameter of DSSF biscuits at both inclusion levels was significantly lower. Thickness also decreased at the 36% DSSF inclusion level, while there was no significant difference in thickness between the control and the 18% DSSF biscuits. The decrease in thickness and diameter in biscuits with DSSF inclusion might be due to the dilution of gluten [17] or increase in fibre content [24] and is in agreement with similar studies on by-product incorporation [10,19]. Cookie diameter is considered a quality indicator and cookies with larger diameters are usually more desirable [25].

There was no significant difference in spread ratio between the control and the 18% DSSF biscuits, while the 36% DSSF biscuits had a significantly higher spread ratio. This might be due to the higher fat content. As explained by Kuchtová, Karovičová, Kohajdová, Minarovičová and Kimličková [10], an increase in fat content leads to an increase in spread ratio, which might be due to the higher fat content in the by-product. Usually the higher the spread ratio of the biscuit, the more desirable it is [26].

DSSF biscuits were harder than the control, which is in contrast to similar studies on by-product incorporation such as apple pomace or grape pomace in biscuits [10,19]. This might be due to the fact that DSSF has a very high protein content compared to other by-products such as apple pomace or grape pomace, which might contribute to hardness. The contribution of protein content to biscuit hardness has been previously reported with whey protein concentrates and defatted soy flour addition [27,28].

Lightness decreased significantly with increasing DSSF inclusion levels. This was expected as DSSF is darker in colour compared to wheat flour, as seen in Table 1. The same pattern can be seen for a* and b*, as these parameters also significantly decreased with increasing DSSF addition, indicating a more intense green and less intense yellow colour in DSSF biscuits compared to the control. As expected, the ΔE, representing the overall difference in colour compared to the control, increased with increasing DSSF addition. These results agree with those from Kuchtová, Karovičová, Kohajdová, Minarovičová and Kimličková [10], Bhat and Hafiza [25] and de Toledo et al. [29], reporting colour alterations in biscuits enriched with by-products.

Table 2. Physical properties of control and defatted sunflower seed flour biscuits.

Parameters		Control	18% DSSF	36% DSSF
Diameter (mm)		55.15 ± 1.26 [a]	54.18 ± 1.03 [b]	53.9 ± 0.91 [b]
Thickness (mm)		8.66 ± 0.40 [a]	8.62 ± 0.27 [a]	8.13 ± 0.25 [b]
Spread ratio		6.38 ± 0.35 [b]	6.29 ± 0.24 [b]	6.64 ± 0.26 [a]
Colour	L*	38.10 ± 0.88 [a]	37.40 ± 0.58 [b]	36.55 ± 0.42 [c]
	a*	6.8 ± 0.26 [a]	4.96 ± 0.26 [b]	3.93 ± 0.28 [c]
	b*	5.16 ± 0.60 [a]	4.49 ± 0.39 [b]	3.87 ± 0.37 [c]
Delta E		-	4.3	11.9
Hardness (N)		40.98 ± 5.44 [b]	50.25 ± 5.53 [a]	53.27 ± 7.10 [a]

Data are expressed as means ±SD on at least 10 biscuits for each of the three batches. Values with the same letter in the same row are not significantly different at $p < 0.05$.

3.3. Effect of Defatted Sunflower Seed Flour on the Proximate Conmposition and Chemical Properties of Biscuits

The proximate composition of control and DSSF biscuits is shown in Table 3. Protein, fat and ash content were significantly higher in DSSF biscuits compared to the control, while the carbohydrate content was lower. There was also a significant difference between DSSF biscuits, with the 36% DSSF biscuits showing the highest protein and ash values between the two, due to the higher DSSF percentage of inclusion. Biscuits with 36% DSSF could be labelled as a "source of protein", because at least 12% of the biscuit calories come from protein [30].

Table 3. Proximate composition of control and defatted sunflower seed flour biscuits.

Parameters	Control	18% DSSF	36% DSSF
Carbohydrate (%)	69.56 [a]	65.16 [b]	61.42 [c]
Fat (%)	17.37 ± 0.5 [b]	18.33 ± 0.1 [a]	18.47 ± 0.4 [a]
Protein (%)	7.98 ± 0.08 [c]	10.80 ± 0.12 [b]	13.61 ± 0.18 [a]
Ash (%)	2.18 ± 0.05 [c]	2.68 ± 0.03 [b]	3.27 ± 0.04 [a]
Estimated calories (Kcal/100 g)	465	467	464
Calories from protein (%)	7	9	12

Data are expressed as means ± SD of duplicate assays. Values with the same letter in the same row are not significantly different at $p < 0.05$.

TPC results (Figure 2A) show that control biscuits had the lowest phenolic content, while DSSF biscuits had significantly higher TPC values. The DPPH and CUPRAC assays (Figure 2B,C) show significant differences among the three recipes, with antioxidant capacity being lowest in the control and then increasing significantly with increasing DSSF inclusion. The higher antioxidant capacity of the 36% DSSF biscuits can be explained by the higher TPC content, since high correlation between antioxidant capacity and phenolic compounds has been reported [31]. Similar results have been reported by Gbenga-Fabusiwa et al. [32], who found that biscuits produced from pigeon pea–wheat flour had higher phenolic content and antioxidant activities compared to those produced with wheat flour only. Aksoylu et al. [33] reported higher TPC in biscuits made with blueberry and grape seeds, while Ajila, Leelavathi and Rao [12] reported that biscuits with mango peel powder had higher DPPH activity.

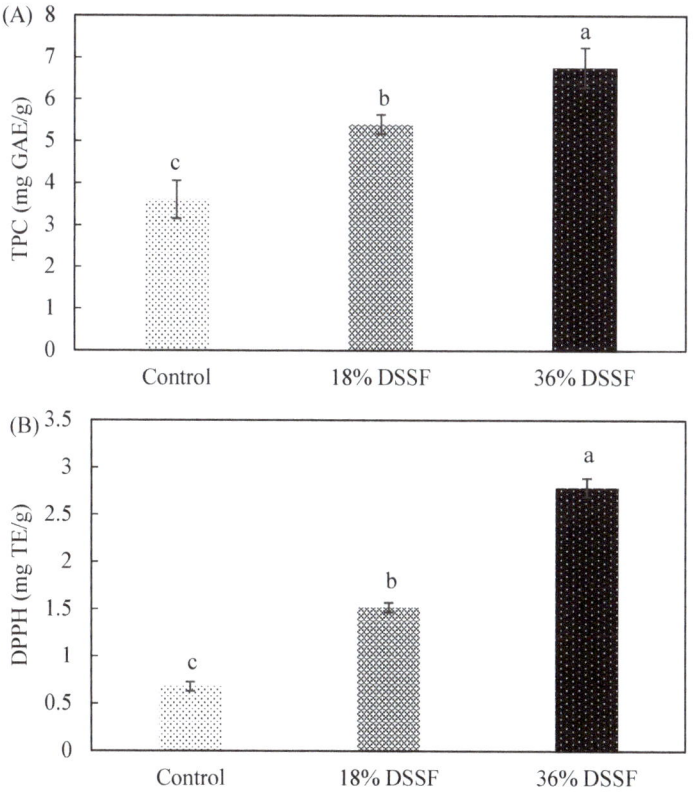

Figure 2. *Cont.*

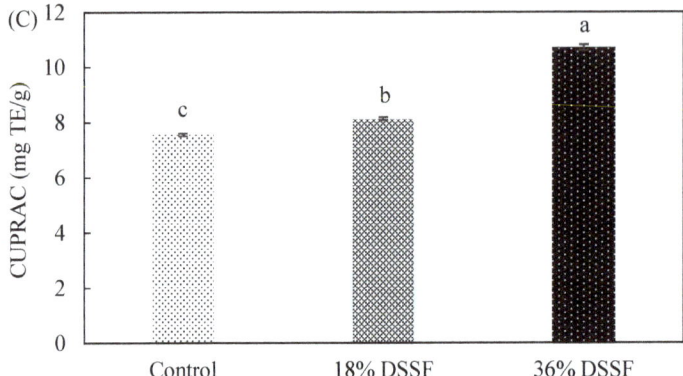

Figure 2. Total phenolic content (TPC) (**A**), DPPH (**B**), and CUPRAC (**C**) values of control and defatted sunflower seed flour (DSSF) biscuits at 18% and 36% inclusion. Data are expressed as means ± SD ($n = 3$). Bars with same letter are not significantly different at $p < 0.05$.

3.4. Effect of Defatted Sunflower Seed Flour on Sensory Properties of Biscuits

The trained panel detected significant differences in eight descriptors of the 29 rated (Figure 3 and Table 4). In terms of appearance, panellists scored DSSF biscuits as more brown than the control, which is in accordance with the instrumental colour test results. Another appearance descriptor that differed was crumb aeration, with the 18% DSSF and control biscuits showing similar crumb aeration, while the 36% DSSF biscuits showed a less aerated crumb. This also corresponds to instrumental measurements, which showed that the 36% DSSF biscuits were significantly less thick compared to the control and 18% DSSF biscuits.

The only aroma descriptor that differed between the samples was burnt aroma. This descriptor scored significantly higher in DSSF biscuits compared to the control, probably due to an acceleration of the Maillard reaction rate in DSSF biscuits [34]. This could be related to the higher amino acid content and the lower sugars in DSSF. Similar results for the descriptor baked flavour were reported by Alongi, Melchior and Anese [19] where apple pomace at 18% and 36% inclusion was added in biscuits.

The only taste descriptor that was significantly different among the three biscuit recipes was off-note. Significantly higher off-note scores were found in the 36% DSSF biscuits, while the control and 18% DSSF biscuits were similar in this parameter. These results could be associated with the bitter and astringent taste of by-products, which is due to the high phenolic content [10,35]. On the other hand, the addition of DSSF even at 36% did not significantly affect the sweet taste descriptor scores, which is a positive finding when compared to the decrease in sweetness reported by Davidov-Pardo et al. [36] in cookies with grape seed extracts.

In terms of texture, 36% DSSF biscuits were significantly harder than 18% DSSF and control biscuits, which concurs with results from instrumental measurements. For the descriptor crumbly, control and 18% DSSF biscuits scored similarly, while 36% DSSF biscuits scored significantly lower, possibly indicating that these biscuits were more compact, less aerated and therefore behaved differently during mastication. Interestingly, the descriptor grainy in relation to biscuit texture was never used by the panellists, while similar studies on incorporation of grape, blueberry and poppy by-products reported grainy textures and rough structures in biscuits [10,33]. The small particle size of DSSF (US mesh 100) might have been beneficial in preventing issues related to graininess.

Finally, differences were perceived on two after effects descriptors, drying and bitter. Again 36% DSSF biscuits scored higher values in these descriptors compared to control and 18% DSSF biscuits. The abundant phenolic content of DSSF could be responsible for observed drying and bitter aftertaste in 36% DSSF biscuits. Upon consumption, the phenolic compounds in DSSF may interact with the glycoproteins in saliva, resulting in less saliva being available to dissolve the biscuits and spread the fat in the mouth [36].

In general, the inclusion of DSSF in biscuits led to sensory changes, more noticeable at high inclusion levels of 36%, but not as much with the lower 18% inclusion. Similar conclusions were reached by Alongi, Melchior and Anese [19]. These authors observed significant changes to the sensory profile of biscuits where wheat flour was replaced with 20% apple pomace, while with a lower concentration of 10% no changes were perceived compared to the control. Similarly, de Toledo, Nunes, da Silva, Spoto and Canniatti-Brazaca [29] found that replacing up to 14% of wheat flour with pineapple, apple and melon by-products did not result in significant sensory differences, while higher inclusion levels did.

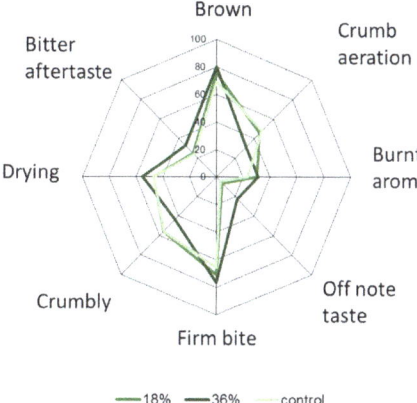

Figure 3. Sensory scores of descriptors that differed significantly between biscuits (control, 18% and 36% defatted sunflower seed flour).

Table 4. Sensory scores of control and defatted sunflower seed flour biscuits.

Parameters	Descriptor	Control	18% DSSF	36% DSSF	p-Value
Appearance	Brown	69.4 [b]	75.6 [a]	80.3 [a]	0.0024
	Crumb aeration	48.2 [a]	45.4 [a]	31.3 [b]	0.0001
Aroma	Burnt	23.6 [b]	30.6 [a]	31.3 [a]	0.0390
Taste and flavour	Off note	4.1 [b]	6.3 [b]	22.3 [a]	0.0162
Mouthfeel	Firm bite	65.5 [b]	70.7 [ab]	76.2 [a]	0.0313
	Crumbly	55.5 [a]	56.0 [a]	43.6 [b]	0.0219
After effects	Drying	46.7 [b]	46.9 [b]	55.0 [a]	0.0053
	Bitter	23.5 [b]	24.8 [b]	32.5 [a]	0.0385

Data are expressed as means of duplicate scoring sessions. Values with the same letter in the same row are not significantly different at $p < 0.05$.

4. Conclusions

This study concluded that where upcycled DSSF was used to replace flour in a short-dough biscuit, the protein content substantially increased, as did the antioxidant capacity and TPC of the biscuits. An 18% replacement of wheat flour with DSSF led to products that were significantly different from the control in only two attributes (brown colour and burnt aroma). The 36% inclusion resulted in biscuits that were significantly less crumbly, less aerated, with a higher off-note, higher drying and bitter after taste compared to both the 18% DSSF and control biscuits. Future work could focus on reformulation aiming to test smaller inclusion levels at 9% and 27% and to test the use of additional ingredients to optimize the recipe.

Author Contributions: Conceptualization, S.G.; Data curation, S.G., E.O., W.X. and K.P.; Formal analysis, S.G., E.O., W.X. and K.P.; Funding acquisition, S.G. and L.M.; Investigation, S.G., E.O., W.X. and K.P.; Methodology, S.G., E.O., W.X. and K.P.; Project administration, S.G.; Resources, S.G. and L.M.; Software, S.G., E.O., W.X., K.P. and L.M.; Supervision, S.G. and K.P.; Validation, S.G., E.O., W.X. and K.P.; Visualization, S.G. and L.M.; Writing—original draft, S.G. and K.P.; Writing—review and editing, S.G., E.O., K.P. and L.M.

Funding: This research received no external funding. It was funded by a Food Pump Priming Awards from the Research Dean for Food at the University of Reading, grant number E3630500.

Acknowledgments: The authors are grateful to the company Planetarians for donating the defatted sunflower seed flour and to Les Crompton and Paul Kirton for their help on protein analysis.

Conflicts of Interest: The authors declare no conflict of interest.

References

1. Yegorov, B.; Turpurova, T.; Sharabaeva, E.; Bondar, Y. Prospects of using by-products of sunflower oil production in compound feed industry. *J. Food Sci. Technol. Ukr.* **2019**, *13*, 106–113. [CrossRef]
2. González-Pérez, S.; Merck, K.B.; Vereijken, J.M.; van Koningsveld, G.A.; Gruppen, H.; Voragen, A.G.J. Isolation and Characterization of Undenatured Chlorogenic Acid Free Sunflower (Helianthus annuus) Proteins. *J. Agric. Food Chem.* **2002**, *50*, 1713–1719. [CrossRef] [PubMed]
3. Wanjari, N.; Waghmare, J. Phenolic and antioxidant potential of sunflower meal. *Adv. Appl. Sci. Res.* **2015**, *6*, 221–229.
4. Francis, G.; Makkar, H.P.S.; Becker, K. Antinutritional factors present in plant-derived alternate fish feed ingredients and their effects in fish. *Aquaculture* **2001**, *199*, 197–227. [CrossRef]
5. Zhang, X.; Han, G.; Jiang, W.; Zhang, Y.; Li, X.; Li, M. Effect of Steam pressure on chemical and structural properties of kenaf fibers during steam explosion process. *BioResources* **2016**, *11*, 10. [CrossRef]
6. Jung, J.Y.; Heo, J.M.; Yang, J.K. Effects of steam-exploded wood as an insoluble dietary fiber source on the performance characteristics of Broilers. *BioResources* **2019**, *14*, 1512–1524.
7. Gu, X.; Dong, W.; He, Y. Detoxification of Rapeseed Meals by Steam Explosion. *J. Am. Oil Chem. Soc.* **2011**, *88*, 1831–1838. [CrossRef]
8. Zhao, Z.M.; Wang, L.; Chen, H.Z. A novel steam explosion sterilization improving solid-state fermentation performance. *Bioresour. Technol.* **2015**, *192*, 547–555. [CrossRef]
9. Planetarians. White Paper. Planetarians: Helping Companies to Find Better Ingredients for People and the Planet. Available online: https://www.planetarians.com/planetarians-technology (accessed on 28 June 2019).
10. Kuchtová, V.; Karovičová, J.; Kohajdová, Z.; Minarovičová, L.; Kimličková, V. Effects of white grape preparation on sensory quality of cookies. *Acta Chim. Slovaca* **2016**, *9*, 84–88. [CrossRef]
11. The Association of Official Analytical Chemists. *AOAC (2000) Office Methods of Analysis*, 17th ed.; AOAC International: Gaithersburg, MD, USA, 2000.
12. Ajila, C.; Leelavathi, K.; Rao, U.P. Improvement of dietary fiber content and antioxidant properties in soft dough biscuits with the incorporation of mango peel powder. *J. Cereal Sci.* **2008**, *48*, 319–326. [CrossRef]
13. Singleton, V.L.; Rossi, J.A. Colorimetry of total phenolics with phosphomolybdic-phosphotungstic acid reagents. *Am. J. Enol. Vitic.* **1965**, *16*, 144–158.
14. Vamanu, E.; Nita, S. Antioxidant capacity and the correlation with major phenolic compounds, anthocyanin, and tocopherol content in various extracts from the wild edible Boletus edulis mushroom. *BioMed Res. Int.* **2013**, *2013*. [CrossRef] [PubMed]
15. Papoutsis, K.; Pristijono, P.; Golding, J.B.; Stathopoulos, C.E.; Bowyer, M.C.; Scarlett, C.J.; Vuong, Q.V. Optimizing a sustainable ultrasound-assisted extraction method for the recovery of polyphenols from lemon by-products: Comparison with hot water and organic solvent extractions. *Eur. Food Res. Technol.* **2018**, *244*, 1353–1365. [CrossRef]
16. Apak, R.; Güçlü, K.; Özyürek, M.; Karademir, S.E. Novel total antioxidant capacity index for dietary polyphenols and vitamins C and E, using their cupric ion reducing capability in the presence of neocuproine: CUPRAC method. *J. Agric. Food Chem.* **2004**, *52*, 7970–7981. [CrossRef] [PubMed]
17. Kohajdová, Z.; Karovičová, J.; Magala, M.; Kuchtová, V. Effect of apple pomace powder addition on farinographic properties of wheat dough and biscuits quality. *Chem. Pap.* **2014**, *68*, 1059–1065. [CrossRef]
18. Sudha, M.L.; Baskaran, V.; Leelavathi, K. Apple pomace as a source of dietary fiber and polyphenols and its effect on the rheological characteristics and cake making. *Food Chem.* **2007**, *104*, 686–692. [CrossRef]

19. Alongi, M.; Melchior, S.; Anese, M. Reducing the glycemic index of short dough biscuits by using apple pomace as a functional ingredient. *LWT* **2019**, *100*, 300–305. [CrossRef]
20. Ratcliff, R.K. *Nutritional Value of Sunflower Meal for Ruminants*; Texas Tech University: Lubbock, TX, USA, 1977.
21. Hidalgo, A.; Brandolini, A.; Čanadanović-Brunet, J.; Ćetković, G.; Šaponjac, V.T. Microencapsulates and extracts from red beetroot pomace modify antioxidant capacity, heat damage and colour of pseudocereals-enriched einkorn water biscuits. *Food Chem.* **2018**, *268*, 40–48. [CrossRef]
22. Lomascolo, A.; Uzan-Boukhris, E.; Sigoillot, J.C.; Fine, F. Rapeseed and sunflower meal: A review on biotechnology status and challenges. *Appl. Microbiol. Biotechnol.* **2012**, *95*, 1105–1114. [CrossRef]
23. Vaher, M.; Matso, K.; Levandi, T.; Helmja, K.; Kaljurand, M. Phenolic compounds and the antioxidant activity of the bran, flour and whole grain of different wheat varieties. *Procedia Chem.* **2010**, *2*, 76–82. [CrossRef]
24. Srivastava, S.; Genitha, T.; Yadav, V. Preparation and quality evaluation of flour and biscuit from sweet potato. *J. Food Process Technol.* **2012**, *3*, 113–118. [CrossRef]
25. Bhat, M.; Hafiza, A. Physico-chemical characteristics of cookies prepared with tomato pomace powder. *J. Food Process. Technol.* **2016**, *7*, 1–4.
26. Chauhan, A.; Saxena, D.; Singh, S. Physical, textural, and sensory characteristics of wheat and amaranth flour blend cookies. *Cogent Food Agric.* **2016**, *2*, 1125773. [CrossRef]
27. Gandhi, A.; Kotwaliwale, N.; Kawalkar, J.; Srivastav, D.; Parihar, V.; Nadh, P.R. Effect of incorporation of defatted soyflour on the quality of sweet biscuits. *J. Food Sci. Technol.* **2001**, *38*, 502–503.
28. Gallagher, E.; Kenny, S.; Arendt, E.K. Impact of dairy protein powders on biscuit quality. *Eur. Food Res. Technol.* **2005**, *221*, 237–243. [CrossRef]
29. De Toledo, N.M.V.; Nunes, L.P.; da Silva, P.P.M.; Spoto, M.H.F.; Canniatti-Brazaca, S.G. Influence of pineapple, apple and melon by-products on cookies: Physicochemical and sensory aspects. *Int. J. food Sci. Technol.* **2017**, *52*, 1185–1192. [CrossRef]
30. EFSA. EU Register on Nutrition and Health Claims. Available online: https://ec.europa.eu/food/safety/labelling_nutrition/claims/nutrition_claims_en (accessed on 28 June 2019).
31. Piluzza, G.; Bullitta, S. Correlations between phenolic content and antioxidant properties in twenty-four plant species of traditional ethnoveterinary use in the Mediterranean area. *Pharm. Biol.* **2011**, *49*, 240–247. [CrossRef]
32. Gbenga-Fabusiwa, F.J.; Oladele, E.P.; Oboh, G.; Adefegha, S.A.; Oshodi, A.A. Polyphenol contents and antioxidants activities of biscuits produced from ginger-enriched pigeon pea–wheat composite flour blends. *J. Food Biochem.* **2018**, *42*, e12526. [CrossRef]
33. Aksoylu, Z.; Çağindi, Ö.; Köse, E. Effects of blueberry, grape seed powder and poppy seed incorporation on physicochemical and sensory properties of biscuit. *J. Food Qual.* **2015**, *38*, 164–174. [CrossRef]
34. Martins, S.; Jongen, W.; van Boekel, M. A review of Maillard reaction in food and implications to kinetic modelling. *Trends Food Sci. Technol.* **2001**, *11*, 364–373. [CrossRef]
35. Naknaen, P.; Itthisoponkul, T.; Sondee, A.; Angsombat, N. Utilization of watermelon rind waste as a potential source of dietary fiber to improve health promoting properties and reduce glycemic index for cookie making. *Food Sci. Biotechnol.* **2016**, *25*, 415–424. [CrossRef] [PubMed]
36. Davidov-Pardo, G.; Moreno, M.; Arozarena, I.; Marín-Arroyo, M.; Bleibaum, R.; Bruhn, C. Sensory and consumer perception of the addition of grape seed extracts in cookies. *J. Food Sci.* **2012**, *77*, S430–S438. [CrossRef] [PubMed]

© 2019 by the authors. Licensee MDPI, Basel, Switzerland. This article is an open access article distributed under the terms and conditions of the Creative Commons Attribution (CC BY) license (http://creativecommons.org/licenses/by/4.0/).

Article

Rheological Properties of Wheat–Flaxseed Composite Flours Assessed by Mixolab and Their Relation to Quality Features

Georgiana Gabriela Codină, Ana Maria Istrate, Ioan Gontariu and Silvia Mironeasa *

Faculty of Food Engineering, Ştefan cel Mare University, 13 Universităţii Street, 720229 Suceava, România
* Correspondence: silviam@fia.usv.ro or silvia_2007_miro@yahoo.com; Tel.: +40-741-985-648

Received: 30 June 2019; Accepted: 5 August 2019; Published: 9 August 2019

Abstract: The effect of adding brown and golden flaxseed variety flours (5%, 10%, 15% and 20% w/w) to wheat flours of different quality for bread-making on Mixolab dough rheological properties and bread quality was studied. The flaxseed–wheat composite flour parameters determined such as fat, protein (PR), ash and carbohydrates (CHS) increased by increasing the level of flaxseed whereas the moisture content (MC) decreased. The Falling Number values (FN) determined for the wheat–flaxseed composite flours increased by increasing the level of flaxseed. Within Mixolab data, greater differences were attributed to the eight parameters analysed: water absorption, dough development time, dough stability and all Mixolab torques during the heating and cooling stages. Also, a general decreased was also recorded for the differences between Mixolab torques which measures the starching speed (C3-2), the enzymatic degradation speed (C4-3) and the starch retrogradation rate (C5-4), whereas the difference which measures the speed of protein weakening due to heat (C1-2) increased. Composite dough behaviour presented a close positive relationship between MC and DT, and FN and PR with the C1-2 at a level of $p < 0.05$. The bread physical and sensory quality was improved up to a level of 10–15% flaxseed flour addition in wheat flour.

Keywords: wheat–flaxseed composite; analytical quality; Mixolab; principal component analysis

1. Introduction

Bread is one of the most consumed food product all over the world. However, the white bread obtained from refined wheat flour (WF) is rather high in carbohydrates and low in proteins, fibre, fat and minerals [1,2]. Therefore, nowadays the actual trend is to improve white bread quality from the nutritional point of view. The addition of flaxseed in bread-making may improve bread quality due to its composition because it is a rich source of essential amino acids, omega 3-fatty acid, dietary fibres, phenolic compounds, e.g., Tobias-Espinoza et al. [3], Oomah [4]. In the world there are many species of flaxseed (*Linum Usitatissimum* L.) varying in colour from brown to light gold [5,6]. The flaxseed varieties do not present significant differences in terms of their chemical composition, but only in terms of the amount of pigments present in the flaxseed, namely, the lower amount of pigments are, the lighter the seed colour is [7]. From the chemical composition point of view, the flaxseed contains about 40–50% fat content, 23–34% protein content, 4% mucilage and 5% ash [8]. It is the leading plant source in the alpha-linolenic acid content (omega-3 fatty acid), being five times higher than in canola oil and walnuts [5]. It is also a rich plant in some amino acids deficient in WF as lysine, valine and tryphtophan [9]. The flaxseed mucilage may be considered a food hydrocolloid due to its composition which consists of a mixture of neutral arabinoxylans and acidic rhamnose-containing polysaccharides [10]. Regarding its mineral content flaxseed is a rich source of K, Mg, Na, Cu, Mn, Zn and Fe [11].

Nowadays, two varieties of flaxseed are known all over the world, namely the golden and the brown one. Of the two varieties, the brown is more widely cultivated than the golden one. However, the golden variety is expected to minimally affect the colour of the final products. The effect of flaxseed flour addition on dough rheological properties was previously studied [12–17]. Some studies showed that dough stability decreased with the increased level of flaxseed addition [12,14] or may increase if high levels are incorporated in WF dough [15–17]. Dough extensibility decreased with the increased level of flaxseed flour addition [13]. Regarding dough behaviour during heating, very few studies have been made for wheat samples in which flaxseed flour was incorporated. However, it seems that flaxseed presents a delay effect on starch gelatinization process [12,13]. Regarding the effect of adding flaxseed to WF, studies have shown that bread quality was generally improved for the samples in which flaxseed flour was incorporated [15,18–20].

The aim of this study was to carry out a complex analysis of the effect of two varieties of flaxseed flour addition to refined, different quality wheat flours on dough rheological properties and bread quality.

Though there have been previous reports on the physical and sensory characteristics of flaxseed-fortified bakery products, there has been a scarcity of reports on its effect on dough mixing and pasting behaviour using a complex device as the Mixolab and of studies between the physico-chemical parameters of wheat–flaxseed composite flours and rheological parameters of these ones.

2. Materials and Methods

2.1. Flour Samples

Two commercial refined WFs with different qualities for bread-making were purchased from S.C. Mopan S.A. Company (Suceava, Romania). The samples were analysed according to the Romanian or international standard methods: gluten deformation and wet gluten according to SR 90/2007, moisture content according to ICC method 110/1, fat content according to ICC 136, protein content according to ICC 105/2, falling number according to ICC 107/1 and ash content according to ICC 104/1. The flaxseed varieties were purchased from SC DECO ITALIA SRL Cluj-Napoca, Romania and they were analysed for their chemical characteristics such as moisture content according to ICC 110/1, protein content according to ICC 105/2, fat content according to ICC 136 and ash content according to ICC 104/1. Carbohydrate content was determined as a difference of mean values: 100 − (the sum of the ash, protein, moisture content and fat) [21].

2.2. Flour Composites

Flours from two different flaxseed varieties (golden and brown) were incorporated in two commercial WFs at different levels (0, 5, 10, 15 and 20%) resulting in a set of 18 samples. The flour composites were mixed in ratios 100:0, 95:5, 90:10, 85:15 and 80:20 (w/w). For these purpose, two different quality WFs for bread-making were used. These were supplemented with each of the two types of flaxseed flour samples: the golden and the brown variety. The flour composites were analysed accordingly as follows; moisture content (ICC 110/1), fat content (ICC 136), protein content (105/2), ash content (ICC 104/1) and falling number (ICC 107/1). Carbohydrate content was determined as a difference of mean values: 100 − (the amount of the protein, moisture content, fat and ash) [21].

2.3. Evaluation of Flour Composite Dough Rheological Properties

The dough mixing and pasting properties of the different wheat/flaxseed flour blends were studied using the Mixolab device (Chopin, Tripetteet Renaud, Paris, France). The composite flours rheological properties were determined according to ICC standard method No. 173. The Mixolab protocol was established as follows; total time to run the analysis 45 min, heating rate 4 °C/min and mixing temperature 30 °C. All the samples were made to optimum hydration level of composite flours in order to achieve the optimum consistency of dough corresponding to the C1 value of 1.1 N·m.

The Mixolab parameters analysed were WA: water absorption (%); DT: dough development time (min); ST: dough stability (min); C2: minimum torque value, corresponding to the initial heating (N·m); C1-2: difference between C1 and C2 peak values (N·m), which measures the speed of protein weakening due to heat: C3, which expressed the starch gelatinization; C4, which expressed the stability of the starch gel formed; C5 torques (N·m), which expressed the starch retrogradation during the cooling stage and C3-2: the difference between C3 and C2 peak values (N·m), which measures the starching speed; C4-3: the difference between C4 and C3 peak values (N·m), which measures the enzymatic degradation speed; and C5-4: the difference between C5 and C4 peak values (N·m), which measures the starch retrogradation rate.

2.4. Bread-making

The bread formulations contained 100 g wheat–flaxseed composite flours (mixed in wheat: flaxseed ratios of 100:0, 95:5, 90:10, 85:15 and 80:20 (w/w)), commercial compressed yeast *Sacharomyces cerevisiae* type (3% flour basis), sodium chloride (2% flour basis) and water up to optimum wheat–flaxseed composite flour hydration capacity. All the ingredients were mixed at the speed of 200 rpm for 15 min in a laboratory mixer (Lancom, Shanghai, China). Then, the dough samples were modelled and placed into loaf pans. These were placed in a fermentation chamber (PL2008, Piron, Italy) for 60 min at 30 °C and 85% relative humidity. Finally, the samples were baked in an oven (PF8004D, Piron, Italy) for 30 min at 180 °C. Bread samples were cooled for 2 h, and then were subjected to physical and sensory analysis.

2.5. Evaluation of Bread Physical Characteristics

Bread physical characteristics (loaf volume, porosity, elasticity) were analysed according to the Romanian standard method SR 91:2007.

2.6. Sensory Evaluation

The bread sensory characteristics were evaluated using semitrained panellists (20 persons). The overall acceptability, appearance, colour, flavour and texture of the samples were evaluated on a nine-point hedonic scale, scoring from one (extremely dislike) to nine (extremely like).

2.7. Statistical Analysis

The statistical analysis of triplicate determinations was performed using the XLSTAT statistical package (free trial version 2016, Addinsoft, Inc., Brooklyn, NY, USA) at a significance level of $p < 0.05$. The data were analysed using variance analysis (ANOVA) and the Tukey test for mean comparison. Principal component analysis (PCA) was used to analyse the intercorrelation between all the variables studied using Statistical Package for Social Science (v.16, SPSS, Chicago, IL, USA).

3. Results and Discussion

3.1. Flour Characteristics

The analytical characteristics of WF of two different qualities—strong quality (F1) and medium quality (F2)—for bread-making according to the Romanian standard SR 877:1996 used in this study are shown in Table 1. The falling number values of the both WFs shows that they have a low α amylase activity due to the fact that they have high FN values (>320 s) [22,23].

Table 1. Quality characteristics of wheat flour (mean value ± standard deviation).

Parameters	Strong Flour F1	Medium Flour F2
Moisture (%)	13.90 ± 0.01	14.50 ± 0.01
Fat (%)	1.7 ± 0.01	1.5 ± 0.01
Protein content (%)	12.2 ± 0.01	12.6± 0.01
Ash content (%)	0.65 ± 0.01	0.65 ± 0.01
Carbohydrates (%)	71.55 ± 0.01	70.75 ± 0.01
Wet gluten (%)	27.50 ± 0.10	34.00 ± 0.20
Gluten deformation index (mm)	3.00 ± 0.18	8.00 ± 0.6
Falling Number index (s)	325.00 ± 2.65	380.00 ± 3.92

The chemical composition of brown and golden flaxseed, determined as a percentage of dried substance, is shown in Table 2. The moisture content of flaxseed samples ranged between 5.6% and 6.2%. Ash content of brown flaxseed was as high as 3.5% and that of golden flaxseed was of 3.41%. The fat content was in the range of 41.12 to 42.25% and protein content was between 19.74% and 20.85%, respectively.

Table 2. Parameters of brown and golden flaxseed flours.

Parameters	Mean Value ± Standard Deviation	
	Brown Flaxseed	Golden Flaxseed
Moisture content (%)	6.2 ± 0.07	5.6 ± 0.04
Fat (%)	42.25 ± 1.15	41.12 ± 1.03
Protein content (%)	19.74 ± 0.46	20.85 ± 0.42
Ash (%)	3.50 ± 0.03	3.41 ± 0.02
Carbohydrates (%)	28.31 ± 0.02	29.02 ± 0.01

The fat content of the flaxseed flours was in agreement with those reported by Ganorkar and Jain [24]. Carbohydrates in brown flaxseed were of 28.31% and of 29.02% for golden flaxseed variety.

3.2. Wheat–Flaxseed Composite Flours Physico-Chemical Characteristics

The physico-chemical characteristics of the wheat–flaxseed composite flours are shown in Tables 3 and 4.

In all cases, the increase in the level of flaxseed flour resulted in the in protein, fat, ash and falling number values and a decreased in moisture and carbohydrates content. This fact was expected due to the high flaxseed content in proteins, fats and ash and its lower amount of moisture and carbohydrates compared to WFs samples. A similar trend of these parameters values for the WF with different levels of flaxseed addition were also reported by Marpalle et al. [20], Codină et al. [13], and Wandersleben et al. [25]. The falling number increased by increasing the level of flaxseed addition more in the case of brown flaxseed variety than in that of the golden one.

It is well known that the falling number value is inversely correlated with α-amylase activity in flours [26] and therefore this trend shows that the flaxseed addition in WF decreased the α-amylase activity in the composite flours.

Table 3. The analysis of the variance of brown (BFs) and golden (GFs) flaxseeds addition to the strong quality flour for bread-making.

Characteristics	Type of Flaxseed		F Ratio	Flaxseeds Doses (%)				F Ratio	Flaxseed Type × Doses
	BFs	GFs		0	5	10	15	20		
MC	13.12 ± 0.56 [a]	13.06 ± 0.60 [b]	13 **	13.90 ± 0.08 [eab]	13.49 ± 0.018 [dab]	13.10 ± 0.03 [ba]	12.69 ± 0.05 [ba]	12.30 ± 0.06 [ab]	1154 ***	2 [ns]
Fat	5.74 ± 2.96 [a]	5.64 ± 2.88 [b]	87 ***	1.70 ± 0.02 [g]	3.65 ± 0.03 [f]	5.68 ± 0.05 [e]	7.69 ± 0.09 [d]	9.69 ± 0.12 [c]	62,448 ***	12 ***
PR	12.93 ± 0.59 [a]	13.07 ± 0.64 [b]	4.4 *	12.22 ± 0.02 [eab]	12.58 ± 0.02 [dab]	12.96 ± 0.36 [cab]	13.41 ± 0.09 [ba]	13.84 ± 0.16 [ab]	74.5 ***	0.4 [ns]
Ash	0.93 ± 0.20 [a]	0.92 ± 0.20 [a]	1.51 [ns]	0.65 ± 0.00 [eab]	0.78 ± 0.02 [dab]	0.92 ± 0.02 [cab]	1.06 ± 0.02 [ba]	1.21 ± 0.02 [ab]	466.38 ***	0.24 [ns]
CHS	67.24 ± 3.29 [a]	67.35 ± 3.10 [a]	0.2 [ns]	71.55 ± 0.02 [eab]	69.60 ± 0.37 [dab]	67.30 ± 0.67 [cab]	64.98 ± 0.45 [ba]	62.99 ± 0.07 [ab]	398.1 ***	0.6 [ns]
FN	385.40 ± 40.39 [a]	360.93 ± 32.38 [b]	502.6 ***	325.00 ± 1.78 [a]	346.50 ± 17.52 [f]	374.33 ± 18.40 [e]	398.000 ± 15.44 [d]	410.50 ± 29.08 [c]	1013.2 ***	32 ***

The means values ± standard deviation in one row followed by different letters differ significantly different at * $p < 0.05$; ** $p < 0.01$; *** $p < 0.001$; ns: nonsignificantly ($p > 0.05$). MC: moisture content (%); Fat: fat content (%); PR: protein content (%); Ash: ash content (%); CHS: carbohydrate content (%); FN: falling number value (s); BFs: the mean values of WFs samples with different doses of brown flaxseed flours addition; GFs: the mean values of WFs samples with different doses of golden flaxseed flours addition.

Table 4. The analysis of the variance of brown (BFs) and golden (GFs) flaxseeds addition to the medium quality flour for bread-making.

Characteristics	Type of Flaxseed		F Ratio	Flaxseeds Doses				F ratio	Flaxseed Type × Doses
	BFs	GFs		0	5	10	15	20		
MC	13.66 ± 0.60 [a]	13.60 ± 0.65 [b]	25 **	14.50 ± 0.008 [g]	14.06 ± 0.018 [f]	13.64 ± 0.03 [e]	13.20 ± 0.05 [d]	12.78 ± 0.06 [c]	2545 **	3 *
Fat	5.56 ± 2.98 [a]	5.46 ± 2.89 [b]	158 **	1.5 ± 0.008 [g]	3.50 ± 0.03 [f]	5.51 ± 0.06 [e]	7.51 ± 0.09 [d]	9.53 ± 0.13 [c]	101,264 **	22 **
PR	13.30 ± 0.51 [a]	13.43 ± 0.59 [b]	48 **	12.60 ± 0.02 [g]	13.03 ± 0.13 [f]	13.36 ± 0.06 [e]	13.74 ± 0.09 [d]	14.13 ± 0.13 [c]	728 **	4 *
Ash	0.93 ± 0.20 [a]	0.92 ± 0.20 [a]	1.41 [ns]	0.65 ± 0.008 [ga]	0.78 ± 0.01 [fa]	0.92 ± 0.01 [ea]	1.06 ± 0.01 [da]	1.21 ± 0.02 [ca]	871.19 **	0.31 [ns]
CHS	66.52 ± 3.10 [a]	66.59 ± 3.05 [b]	79 **	70.73 ± 0.01 [g]	68.67 ± 0.02 [f]	66.55 ± 0.04 [e]	64.46 ± 0.06 [c]	62.34 ± 0.07 [c]	148,546 **	11 **
FN	440.06 ± 20.26 [a]	412.00 ± 29.60 [b]	1738 **	380.33 ± 2.25 [g]	391.83 ± 5.87 [f]	420.00 ± 16.49 [e]	454.50 ± 25.75 [d]	483.50 ± 29.08 [c]	3268 **	230 **

The means values ± standard deviation in one row followed by different letters differ significantly different at * $p < 0.05$; ** $p < 0.01$; *** $p < 0.001$; ns: nonsignificantly ($p > 0.05$). MC: moisture content (%); Fat: fat content (%); PR: protein content (%); Ash: ash content (%); CHS: carbohydrate content (%); FN: falling number value (s); BFs: the mean values of WFs samples with different doses of brown flaxseed flours addition; GFs: the mean values of WFs samples with different doses of golden flaxseed flour addition.

3.3. Influence of Flaxseed on Mixolab Dough Rheological Properties

Incorporation of flaxseed from different varieties, brown and golden addition at 0, 5, 10, 15 and 20% in two categories of WF, namely strong (F1) and medium (F2), for bread-making, showed significant differences ($p < 0.001$) in terms of water absorption values. As it may be seen, the addition of brown flaxseed (BFs) and golden flaxseed (GFs), respectively in WF F1 and F2, decreased the water absorption (CH) values (Figure 1a). The lowest decreased in water absorption was found when the brown flaxseed was added in WF F2, from the level of 10% to 15% (55.2–54.6%). The decrease in water absorption in the case of golden flaxseed addition in F2 is in the same trend as the decrease in water absorption in F1. Similar effects on water absorption were observed by Roozegar et al. [16,17] and Codină et al. [12] when brown flaxseed or golden flaxseed were added, respectively. Kundu et al. [27] reported that the difference in water absorption is mainly caused by the gluten dilution, which needs less hydration, and therefore the wheat–flaxseed composite flours require lower amounts of water in the dough system in order to obtain the optimum consistency. As compared to the control samples (the sample without flaxseed flour addition), water absorption decreased in all the mixes made from wheat flour in which the golden or brown flaxseeds were incorporated. The lowest value for water absorption was obtained in the case when 20% level of brown flaxseed was incorporated in the flour of a medium quality for bread-making which decreased by 3.87% as compared to the control sample.

Compared to the wheat flour sample without flaxseed addition, highly significant effects ($p < 0.001$) were noticed in relation to dough development time (DT) values. The highest decreased in DT was observed for the sample with 20% flaxseed addition levels for both BFs and GFs varieties incorporated in F1 and F2, respectively (Figure 1b). These decreases may be due to gluten dilution in the dough system by flaxseed addition. Therefore, the amount of free water will increase, leading to a DT decrease due to the fact that in a dough system the highest amount of water is absorbed by starch and gluten [28] which will trigger a lower amount due to flaxseed addition in wheat flour.

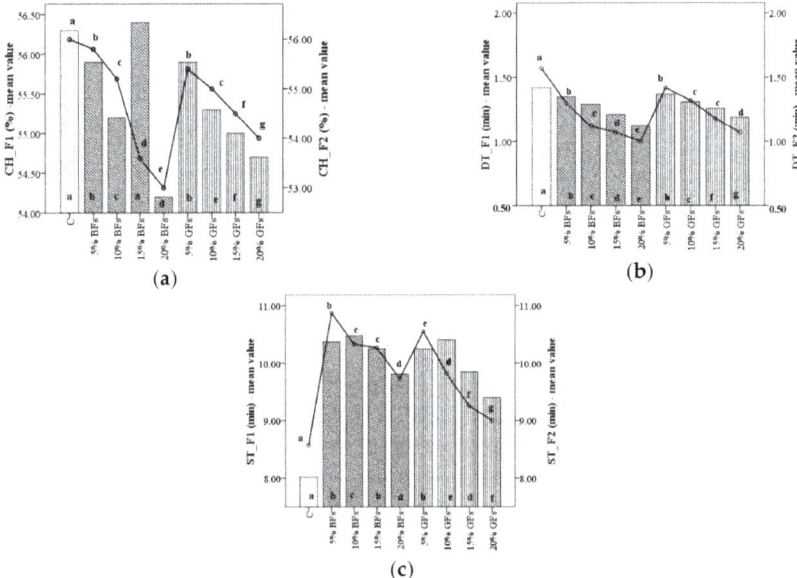

Figure 1. Mixolab parameters during mixing of WF (C) and flaxseed–wheat composite flours formulated by adding different levels (5,10,15 and 20%) of brown flaxseed (BFs) and golden flaxseed flour (GFs), respectively, in two types of flour—F1 and F2: (**a**) CH: water absorption; (**b**) DT: development time; and (**c**) ST: stability. Different letters indicate significant differences ($p < 0.05$) between samples.

Dough stability, which stands for dough strength, significantly decreased (Figure 1c) from 10.37 to 9.81 min and from 10.25 to 9.40 min when brown flaxseed and golden flaxseed were added in WF F1 at a level from 5% to 20%. A similar decrease in ST was noticed and in the case of F2 (from 10.87 to 9.73 min) when BFs was added and from 10.55 to 9.00 min when GFs was incorporated in WF, at the same levels of addition (5–20%). Similar results were reported by Pourabedin et al. [14], Meral and Dogan [15] and Roozegar et al. [16,17] for the addition of flaxseed in WF. However, as compared to the control sample, dough stability values gradually increased in mixes with the increase of levels addition from 0 to 10% for BFs and GFs, respectively, in WF F1, whereas for the WF F2 the ST values increased only for the levels of BFs and GFs up to 5% addition in wheat flour. This may likely be due to the interaction between polysaccharides (especially gums) and proteins in flaxseed–wheat composite flour as reported earlier by Rojas et al. [29]. Also, an increase of dough stability values to a higher level of flaxseed addition for the F1 flour than for the F2 one may be attributed to the WF quality. The F1 flour is of a strong quality for bread-making, which indicates that it can develop stronger and elastic dough than F2 flour which is of a medium quality. This indicates that F1 flour can sustain for a longer period of time higher wheat–flaxseed composite flour dough stability during mixing, compared to F2 flour. By comparing the obtained values for ST one can noticed that stability increased to a greater extent in the case of BFs addition in F1 from 8.02 min for the control sample (C) to 10.37 min for the sample with 5% BFs incorporation and from 8.57 min for the C to 10.87 min for the same level of BFs incorporation in F2. However, for a high level of flaxseed flour addition, a slight decrease in dough stability for both flours in which flaxseeds were incorporated due to gluten dilution was noticed because flaxseed is non-gluten flour.

The effect of incorporation of BFs and GFs flours at varying levels on dough C2 torque represents the protein weakening (C2) as illustrated in Figure 2a. Its values decreased with the increased level of flaxseed addition more in the case of brown variety than in the case when the golden one was used. The lowest values for C2 torque were recorded for the F2 flour with a decreased of 33.3% to a level of 20% brown flaxseed addition as compared to the control sample.

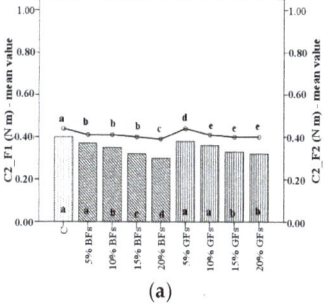
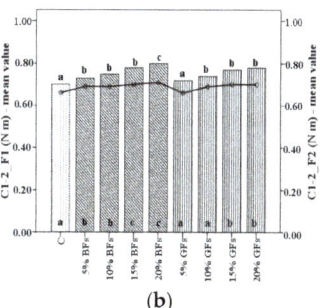

Figure 2. C2 torque (**a**) and difference between torques C1 and C2 (C1-2) (**b**) of WF (C) and flaxseed–wheat composite flours formulated by adding different levels (5%,10%,15% and 20%) of brown flaxseed (BFs) and golden flaxseed flour (GFs) in two types of flour—F1 and F2. Different letters indicate significant differences ($p < 0.05$) between samples.

A similar trend (Figure 2b) may be seen, and in the case of the difference between the peak C2 and C1 values (C2-1), which measures the speed of protein weakening due to heat, obviously increased with the increased level of flaxseed addition. An increase in the C2-1 values and a decrease in the value of C2 together with the increase in the flaxseed addition are due to the protein network structure. By flaxseed flour addition, proteins become less compact, a fact that favours the enzymatic attaching points, leading to an increase in the speed of protein weakening due to heat (C2-1) and a decrease in C2. Therefore, we may conclude according to the data obtained that by flaxseed addition the protein network becomes weaker under the effect of temperature increase.

When dough is heated above 60 °C the Mixolab begins to record the pasting properties of dough the C3 torque and the difference between the C3 and C2 peak values (C3-2) being associated with the starch gelatinization process. In general, both parameters values decreased (Figure 3a,b) with the increased level of flaxseed addition in the case of C3 values, this decrease is highly significant ($p < 0.001$).

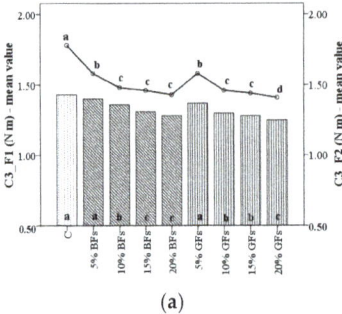
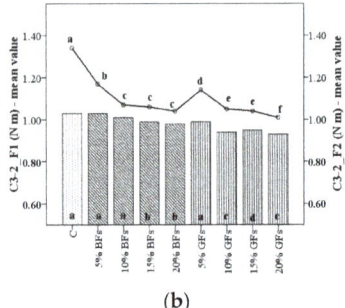

(a) (b)

Figure 3. C3 torque (a) and difference between torques C3 and C2 (C3-2) (b) of WF (C) and flaxseed–wheat composite flours formulated by adding different levels (5%,10%,15% and 20%) of brown flaxseed (BFs) and golden flaxseed flour (GFs) in two types of flour—F1 and F2. Different letters indicate significant differences ($p < 0.05$) between samples.

The decrease in C3 values is higher when the GF variety is added in WF dough with 26.2% for F1 and 14.4% for F2 for 20% level addition as compared to the control sample, probably due to starch dilution and the high content of fat and polysaccharides in flaxseed flour [12,14,29]. It is well known that the starch gelatinization process is influenced by the amylase-lipid complex formation, the amount of amylose leaching, the competition for free water between leached amylose and ungelatinised granules remained [30]. A decrease in C3 might be due to less swelling of the starch granules from WF in the presence of flaxseed flour of whose compounds may interact with amylose. Also, it is possible that some compounds from flaxseed to compete with starch to absorb water during the starch gelatinization process fact that will create difficulties for starch to gelatinize.

The C4 torque values corresponding to the hot starch stability paste decreased with the increased level of flaxseed addition (Figure 4a); more in the case of F2 with 17.14% when GFs were incorporated and with 13.1% when BFs were added in WF to a level of 20%. This effect may be attributed to the lower amylase activity in the wheat–flaxseed composite flour which slows gelatinization process and due to starch dilution, taking into account that flaxseed flour contains low amount of starch [12]. The amount of carbohydrates is less in the case of BFs than GFs and therefore the starch dilution of wheat–flaxseed composite flour is higher when GFs were added than in the case when BFs were incorporated in WF. Also, the stability of the starch gel formed is influenced by starch composition. Flaxseed flour addition in WF may have some interactions between starch and some compounds from the flaxseed flour. For example, the polysaccharides from the flaxseed content probably in a higher amount in BFs than in the case of GFs binds through the hydrogen bonds the water from the dough system, leading to a decrease of available water for the starch granules. Also, the high content of fat from the flaxseed flour may form insoluble complex with amylose leading to a decrease of the hot starch stability paste [31].

The difference between the C4 and C3 peak values (C4-3) did not vary in a significant way with the increased level of flaxseed addition. However, all the C4-3 presented lower values in the samples with flaxseed addition. This fact is explainable since the C4-3 values correspond to the rate of amylases hydrolysis on WF starch. Since the flaxseed flours did not bring amylases in dough system (as we can see from the falling number values) these parameters values decreased as compared to the sample when no flaxseeds were added. The starch retrogradation during the cooling period of the Mixolab device represented by the values of C5 torque and the difference between the C5 and C4 peaks (C5-4) decreased with the increased level of flaxseed addition (Figure 4b,c). This fact shows an anti-staling effect that flaxseed may have on bread quality more in the case of the golden variety than in the case of the brown one. An extent of bread freshness by flaxseed addition has also been reported by Khorshid et al. [19].

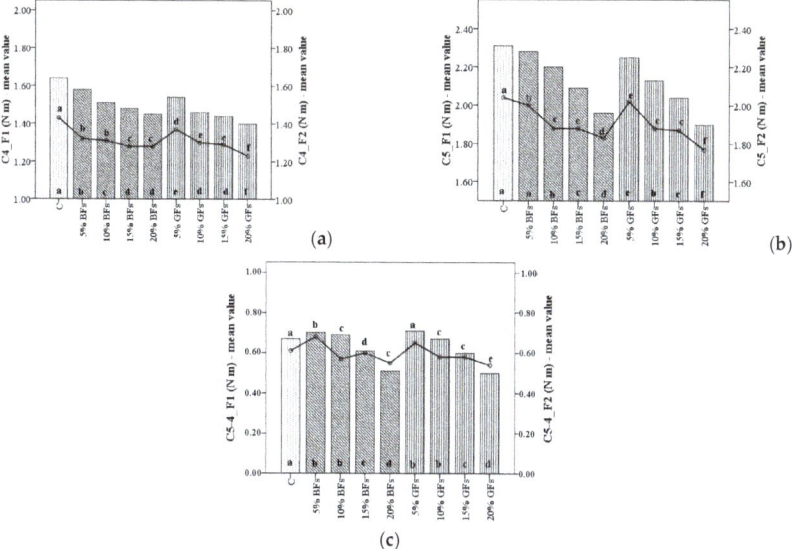

Figure 4. C4 torque (**a**), C5 torque (**b**) and difference between torques C5 and C4 (C5-4) (**c**) of WF (C) and flaxseed–wheat composite flours formulated by adding different levels (5%,10%,15% and 20%) of brown flaxseed (BFs) and golden flaxseed flour (GFs) in two types of flour, F1 and F2. Different letters indicate significant differences ($p < 0.05$) between samples.

The decrease in the C5 values was higher when flaxseed was incorporated in F2 with 21.57% when GFs were added and 17.85% when BFs were added at a substitution level of 20% flaxseed in WF. This fact may be due to the high content of fat and other compounds in the flaxseed flour like polysaccharides that interact with gluten and starch in the dough system hindering a less starch retrogradation [14,19]. This fact is mainly attributed to the interaction between lipids from flaxseed flour with starch especially with amylose during the baking process. The complex formed between amylose and lipids is insoluble in water. In this form, amylose cannot leach out of starch granules. Thus, it decreased the amount of free amylase capable to leach out of gelatinized starch. As a consequence, it decreased its capacity to form intermolecular association during the cooling stage. On the other hand, the amylose remains inside the starch granules in a higher amount. This is due to the lipids presence, which creates difficulties between the associations of amylopectin molecules. Consequently, the starch retrogradation process is delayed [32]. Also, the polysaccharides from the flaxseed flours, which are probably in a lower amount in GFs than in BFs, may form intermolecular associations with leached amylose molecules during pasting which prevents starch retrogradation.

3.4. Correlation Analysis of the Evaluated Parameters for the Wheat–Flaxseed Composite Flours

The PCA was performed on wheat–flaxseed composite flours' characteristics (moisture content (MC), fat content (Fat), protein content (PR), ash content (Ash), carbohydrates content (CHS), falling number (FN)) and dough rheological properties assessed by the Mixolab device (C2, C3, C4 and C5 torques; the difference between the C1 and C2 and peak values (C1-2); the difference between the C3 and C2 and peak values (C3-2); the difference between the C4 and C3 and peak values (C4-3); and the difference between the C5 and C4 and peak values (C5-4)) for the all 18 samples, which were analysed in this study, shown in Figure 5. The results obtained showed that all the variables used in order to perform PCA can be reduced to two principal components (PCs), 56.03% by PC1 and 26.37% by PC2. In PCA, factors extracted are retained if they have an eigenvalue >1 because they provide a lot more information than the initial variables, the first two components explaining 82.40% of the total variance. The plot of PC1 vs. PC2 plot shows along the PC1 axis, a close relationship between the starch pasting properties (C3, C3-2 and C4-3) and between wheat–flaxseed composite flours characteristics (FN, PR, Ash, Fat and MC and CH). It could be noticed that the Mixolab parameters related to dough rheological properties during heating (C1-2, C2, C3, C3-2, C4 and C4-3) were included in PC2 along with the starch retrogradation values at the cooling stage (C5 and C5-4). The influence of ST values was low, since their loadings for PC1 was close to null. Between wheat–flaxseed composite flours characteristics and Mixolab dough rheological properties a close relationship between MC and DT ($r = 0.838$) was noticed, respectively, C5 ($r = 0.919$) at a level of $p < 0.01$. FN was closely associated with the C1-2 value ($r = 0.903$) and inversely correlated to C2 ($r = -0.910$), C3-2 ($r = -0.689$), C3 ($r = -0.808$) at a level of $p < 0.01$. The positive correlation between FN and C1-2 may be due to the fact that the flours used in the analysis had a low α amylase activity and probably a low proteolytic activity.

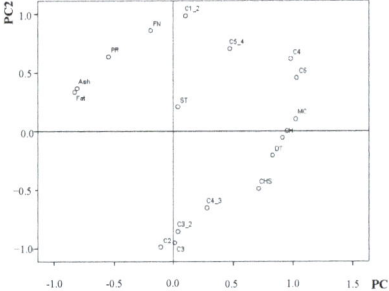

Figure 5. Loading plot of the first two principal components based on physicochemical and rheological properties of the wheat–flaxseed composite flours samples: MC: moisture content; Fat: fat content; PR: protein content; Ash: ash content; CHS: carbohydrates content; WA: water absorption; DT: development time; ST: stability; C2, C3, C4, and C5—Mixolab torques, C1-2, C3-2, C4-3, and C5-4 difference between Mixolab peak values C1 and C2, C3 and C2, C4 and C3 and C5 and C4.

Since C1-2 expresses the speed of protein weakening due to heat, this speed is influenced by the proteolytic activity, and since the flaxseed addition did not improve α amylase activity, it is also probable that it did not to improve the proteolytic activity either. A negative correlation between FN and Mixolab values C3 and C3-2 is also explainable since both parameters are related to starch gelatinization process. The FN is a measure of the α amylase activity, its value being inversely correlated with the α amylase amount in WF and therefore high values of FN show low α amylase activity in dough system [33,34] fact that will negatively influence the starch gelatinization process. Protein content of the samples was significantly positively correlated with C1-2 value ($r = 0.752$) and negatively correlated with C2 ($r = -0.754$), C3-2 ($r = -0.751$), C3 ($r = -0.812$), C4-3 ($r = -0.715$) at a level of $p < 0.01$. Although flaxseeds have higher protein content, these are non-gluten, and therefore are weaker. Also, the Mixolab values related to dough pasting properties decreased due to the fact that by increasing the level of proteins in dough system the starch content decreased [35].

3.5. Influence of Flaxseed on Bread Physical Quality Characteristics

The physical characteristics of the bread samples with different flaxseed content are shown in Tables 5 and 6.

Table 5. Analysis of variance of the influence of flaxseeds from the brown variety (BFs) and golden variety (GFs) addition on physical characteristics of bread obtained from strong quality flour for bread-making.

Characteristics	Type of Flaxseed		F Ratio	Flaxseeds Doses (%)					F-Ratio	Flaxseed Type × Doses
	BFs	GFs		0	5	10	15	20		
Loaf specific volume (cm³/100 g)	290.97 ± 27.56 [a]	294.11 ± 13.80 [b]	74 **	277.49 ± 0.89 [e]	295.44 ± 14.59 [d]	317.61 ± 16.24 [c]	296.96 ± 18.56 [b]	275.22 ± 20.84 [a]	1774 **	1548 **
Porosity (%)	84.30 ± 1.16 [a]	85.84 ± 2.67 [b]	17.8 **	83.50 ± 0.89 [e]	84.90 ± 1.17 [d]	86.50 ± 1.25 [c]	87.05 ± 3.03 [b]	83.40 ± 0.95 [a]	16.9 **	7.9 **
Elasticity (%)	85.00 ± 1.09 [a]	86.06 ± 2.31 [b]	8.4 *	84.20 ± 0.90 [e]	85.25 ± 0.97 [d]	86.10 ± 0.92 [c]	87.77 ± 2.49 [b]	84.32 ± 0.89 [a]	13.0 **	4.9 *

The means values ± standard deviation in one row followed by different letters differ significantly different at * $p < 0.01$; ** $p < 0.001$.

Table 6. Analysis of variance of the influence of flaxseeds from the brown variety (BFs) and golden variety (GFs) addition on the physical characteristics of bread obtained from a medium quality flour for bread-making.

Characteristics	Type of Flaxseed		F Ratio	Flaxseeds Doses (%)					F-Ratio	Flaxseeds Type × Doses
	BFs	GFs		0	5	10	15	20		
Loaf specific volume (cm³/100 g)	351.19 ± 18.34 [a]	373.58 ± 44.24 [c]	3760 *	332.15 ± 0.89 [e]	347.84 ± 8.62 [d]	389.41 ± 10.10 [c]	403.80 ± 45.09 [b]	338.73 ± 2.57 [a]	6196 *	1827 *
Porosity (%)	84.30 ± 2.19 [a]	85.04 ± 1.81 [a]	4.1 ns	83.50 ± 0.89 [e]	85.63 ± 0.89 [d]	87.11 ± 0.91 [c]	84.97 ± 1.19 [b]	82.13 ± 1.38 [a]	22.2 **	1.2 ns
Elasticity (%)	84.99 ± 1.83 [a]	85.66 ± 1.65 [a]	3.3 ns	84.20 ± 0.90 [e]	85.95 ± 0.97 [d]	86.71 ± 1.00 [c]	86.30 ± 1.96 [b]	83.48 ± 1.26 [a]	11.9 **	4.5 *

The means values ± standard deviation in one row followed by different letters differ significantly different at * $p < 0.01$; ** $p < 0.001$; ns: nonsignificantly ($p > 0.05$).

For all bread samples in which GFs were incorporated in the loaf volume, porosity and elasticity increased up to a level of 15% flaxseed addition and then decreased. Also, when BFs was added in WF, all the bread physical characteristics increased up to a level of 10% flaxseed addition and then decreased. An increase of the bread physical characteristics up to a level of 10–15% flaxseed addition may be due to the high amount of lipids from the flaxseed flour. This lipid effect on bread physical characteristics may be due to its presence in the liquid film which surrounds the gas cells. During baking, bread is not only gluten continuous but also gas continuous during the fact that gas cell opening occurs. This is one of the reasons why the bread does not collapse when gases are lost during baking and bread cooling [33]. The efficiency with which gas cell integrity was maintained is connected with the amount of lipids founds in the film which surrounds them and its semicrystalline organisation. The lipids are absorbed to the interface between gases and water forming a physical barrier in coalescence of the carbon dioxide bubbles. This leads to products with higher loaf volume, more fine and uniform porosity compared to the products without lipids addition. However, when high amount of flaxseed was added in WF, the bread physical characteristics begin to decrease, probably due to the gluten dilution effect. The flaxseed addition up to a certain level led to an increase in the bread physical characteristics. This was reported by [15,35,36].

As we can see from the Tables 5 and 6 the physical values for the bread samples obtained from the medium quality flour are higher than the values for the bread obtained from the strong quality flour. This is probably due to the fact that F1 flour is a strong one and makes bread hardly extensible which will affects the growth of the bread. This will lead to products with low loaf volume. The F2 flour is of medium quality for bread-making with a good elasticity and extensibility. Also, it presents a high amount of gluten content which will facilitate a better holding for all dough components. This fact will improve the dough gas retention capacity during the bread-making process, which will lead to an increase of dough volume. As a consequence, the bread volume will increase as well [28].

3.6. Influence of Flaxseed on Bread Sensory Results

The results of sensory tests for the overall acceptability, appearance, colour, flavour and texture are shown in Figures 6 and 7.

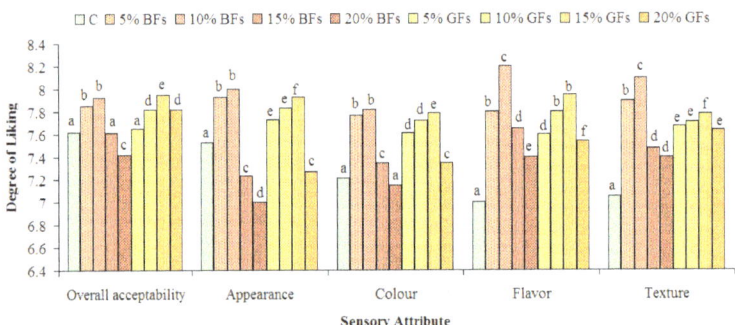

Figure 6. Effect of brown flaxseed (BFs) and golden flaxseed (GFs) at different addition levels: 0% (C), 5%, 10%, 15% and 20% on the sensory attributes of bread from a WF of a strong quality for bread-making. Means with different letters indicate significant difference among treatments ($p < 0.05$).

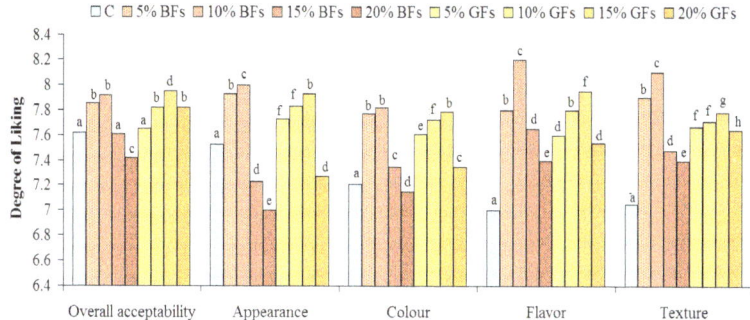

Figure 7. Effect of brown flaxseed (BFs) and golden flaxseed (GFs) at different addition levels: 0% (C), 5%, 10%, 15% and 20% on the sensory attributes of bread from a WF of a good quality for bread-making. Means with different superscripts indicate significant difference among treatments ($p < 0.05$).

The overall acceptability was concluded to be the best for bread samples in which flaxseed was incorporated up to 10%. No significant differences were found for samples with 5 and 10% BFs addition ($p<0.05$). Regarding the appearance and the colour evaluation, these parameters presented higher degree of liking for all bread samples up to a 10% GFs level. Samples with GFs received higher scores due to their more appealing, yellowish colour. The scores for flavour and texture were the highest for the samples in which 10% of BFs were added and for the samples in which 15% GFs were incorporated in WF. For all the sensory parameters evaluated significant differences were found between the control samples and the samples in which flaxseed flour was incorporated.

4. Conclusions

Physico-chemical and rheological properties of the composite flours varied significantly with the increased level of flaxseed addition. The partial substitution of wheat flour with flaxseed flours significantly increased ($p < 0.001$) the amounts of fats, proteins and carbohydrates to 9.69%, 13.84% and 62.99%, respectively, in the case of the wheat flour of a strong quality for bread-making, and to 9.53%, 14.13% and 62.34%, respectively, in the case of wheat flour of a medium quality for bread-making. It seems that flaxseed addition decreased the α-amylase activity in wheat flour since the falling number value increased with the increased level of flaxseed. The Mixolab results showed that water absorption, dough development time, protein weakening peak, starch gelatinization, starch hot-gel stability and retrogradation were significantly ($p < 0.05$) reduced as flaxseed level addition became higher. Stability was significantly ($p < 0.05$) increased up to a level of 5–10% flaxseed addition, after this level its values decreased but to a higher value than that of the control sample.

The graphic representation by PCA provides intuitive and quantitative classification of physico-chemical and Mixolab multidimensional data of wheat flours with different levels of flaxseed addition. The biplots presentation also shows good correlation between physico-chemical and rheological parameters measured by Mixolab device which greatly enhances the ability of understanding how the dough rheological properties provided by the Mixolab data can be affected by the physico-chemical parameters of composite flours. According to the bread quality parameters evaluated, a partial replacement of up to 10–15% flaxseed flour is possible in order to produce bread of a good quality.

Author Contributions: Conceptualization, G.G.C. and S.M.; Methodology, A.M.I. and I.G.; Formal Analysis, G.G.C., A.M.I., I.G. and S.M.; Investigation, A.M.I. and I.G.; Resources, G.G.C. and S.M.; Writing—Original Draft Preparation, G.G.C. and S.M.; Writing—Review and Editing, G.G.C. and S.M.; Project Administration, G.G.C. and S.M.; Funding Acquisition, G.G.C. and S.M.

Funding: This work was supported from contract No. 18PFE/16.10.2018 funded by Ministry of Research and Innovation within Program 1—Development of National Research and Development System, Subprogram 1.2—Institutional Performance—RDI excellence funding projects.

Conflicts of Interest: The authors declare no conflict of interest.

References

1. Bazhay-Zhezherun, S.; Simakhina, G.; Bereza-Kindzerska, L.; Naumenko, N. Qualitative indicators of grain flakes of functional purpose. *Ukr. Food J.* **2019**, *8*, 7–17. [CrossRef]
2. Dewettinck, K.; Van Bockstaele, F.; Kühne, B.; Van de Walle, D.; Courtens, T.M.; Gellynck, X. Nutritional value of bread: Influence of processing, food interaction and consumer perception. *J. Cereal Sci.* **2008**, *48*, 243–257. [CrossRef]
3. Tobias-Espinoza, J.L.; Amaya-Guerra, C.A.; Quintero-Ramos, A.; Pérez-Carrillo, E.; Núñez-González, M.A.; Martínez-Bustos, F.; Meléndez-Pizarro, C.O.; Báez-González, J.C.; Ortega-Gutiérrez, J.A. Effects of the addition of flaxseed and amaranth on the physicochemical and functional properties of instant-extruded products. *Foods* **2019**, *8*, 183. [CrossRef] [PubMed]
4. Oomah, B.D. Flaxseed as a functional food source. *J. Food Sci. Technol.* **2001**, *81*, 889–894. [CrossRef]
5. Oliveira, M.L.; Arthur, V.; Polesi, L.F.; Silva, L.C.; Oliveira, A.L. Evaluation of production and gamma radiation effects in pasta enriched with brown flaxseed bagasse (*Linum Usitatissimum* L.). *Eur. Int. J. Sci. Technol.* **2014**, *3*, 232–238.
6. Kajla, P.; Sharma, A.; Sood, D.R. Flaxseed—A potential functional food source. *J. Food Sci. Technol.* **2015**, *52*, 1857–1871. [CrossRef] [PubMed]
7. Ganorkar, P.M.; Jain, R.K. Flaxseed—A nutritional punch. *Int. Food Res. J.* **2013**, *20*, 519–525.
8. Toure, A.; Xueming, X. Flaxseed lignans: Source, biosynthesis, metabolism, antioxidant activity, bio-active components, and health benefits. *Compr. Rev. Food Sci. Food Saf.* **2010**, *9*, 261–269. [CrossRef]
9. Kaur, M.; Kaur, R.; Gill, B.S. Mineral and amino acid contents of different flaxseed cultivars in relation to its selected functional properties. *J. Food Meas. Charact.* **2017**, *11*, 500–511. [CrossRef]
10. Cui, W.; Mazza, G.; Oomah, B.D.; Biliaderis, C.G. Optimization of an aqueous extraction process for flaxseed gum by response surface methodology. *LWT Food Sci. Technol.* **1994**, *27*, 363–369. [CrossRef]
11. Goyal, A.; Sharma, V.; Upadhyay, N.; Gill, S.; Sihag, M. Flax and flaxseed oil: An ancient medicine & modern functional food. *J. Food Sci. Technol.* **2014**, *51*, 1633–1653.
12. Codină, G.G.; Mironeasa, S. Use of response surface methodology to investigate the effects of brown and golden flaxseed on wheat flour dough microstructure and rheological properties. *J. Food Sci. Technol.* **2016**, *53*, 4149–4158. [CrossRef] [PubMed]
13. Codină, G.G.; Arghire, C.; Rusu, M.; Oroian, M.A.; Sănduleac, E.T. Influence of two varieties of flaxseed flour addition on wheat flour dough rheological properties. *Ann. Univ. Dunarea Jos Galati Fascicle VI Food Technol.* **2017**, *41*, 115–126.
14. Pourabedin, M.; Aarabi, A.; Rahbaran, S. Effect of flaxseed flour on rheological properties, staling and total phenol of Iranian toast. *J. Cereal Sci.* **2017**, *76*, 173–178. [CrossRef]
15. Meral, R.; Dogan, I.S. Quality and antioxidant activity of bread fortified with flaxseed. *Ital. J. Food Sci.* **2013**, *25*, 51–56.
16. Roozegar, M.H.; Shahedi, M.; Hamdami, N. Production and rheological and sensory evaluation of Taftoon bread containing flaxseed. *Iran. J. Food Sci. Technol.* **2015**, *12*, 231–244.
17. Roozegar, M.H.; Shahedi, M.; Keramet, J.; Hamdami, N.; Roshanak, S. Effect of coated and uncoated ground flaxseed addition on rheological, physical and sensory properties of Taftoon bread. *J. Food Sci. Technol.* **2015**, *52*, 5102–5110. [CrossRef]
18. De Lamo, B.; Gómez, M. Bread enrichment with oliseeds. A review. *Foods* **2018**, *7*, 191. [CrossRef]
19. Khorshid, A.M.; Assem, N.H.; Abd, E.M.; Fahim, J.S. Utilization off laxseeds in improving bread quality. *Egypt. J. Agric. Res.* **2011**, *89*, 241–250.
20. Marpalle, P.; Sonawane, S.K.; Arya, S.S. Effect of flaxseed flour addition on physicochemical and sensory properties of functional bread. *LWT* **2014**, *58*, 614–619. [CrossRef]
21. Mironeasa, S.; Codină, G.G.; Mironeasa, C. Optimization of wheat-grape seed composite flour to improve alpha-amylase activity and dough rheological behavior. *Int. J. Food Prop.* **2016**, *19*, 859–872. [CrossRef]
22. Codină, G.G.; Mironeasa, S.; Mironeasa, C.; Popa, C.N.; Tamba-Berehoiu, R. Wheat flour dough Alveograph characteristics predicted by Mixolab regression models. *J. Sci. Food Agric.* **2012**, *92*, 638–644. [CrossRef] [PubMed]

23. Ji, T.; Penning, B.; Baik, B.K. Pre-harvest sprouting resistance of soft winter wheat varieties and associated grain characteristics. *J. Cereal Sci.* **2018**, *83*, 110–115. [CrossRef]
24. Ganorkar, P.M.; Jain, R.K. Effect of flaxseed incorporation on physical, sensorial, textural and chemical attributes of cookies. *Int. Food Res. J.* **2014**, *21*, 1515–1521.
25. Wandersleben Morales, T.E.; Burgos-Díaz, C.; Barahona, T.; Labra, E.; Rubilar, M.; Salvo-Garrido, H. Enhancement of functional and nutritional properties of bread using a mix of natural ingredients from novel varieties of flaxseed and lupine. *LWT* **2018**, *91*, 48–54. [CrossRef]
26. Struyf, N.; Verspreet, J.; Courtin, C.M. The effect of amylolytic activity and substrate availability on sugar release in non-yeasted dough. *J. Cereal Sci.* **2016**, *69*, 111–118. [CrossRef]
27. Kundu, H.; Grewal, R.B.; Goyal, A.; Upadhyay, N.; Prakash, S. Effect of incorporation of pumpkin (*Cucurbitamoshchata*) powder and guar gum on the rheological properties of wheatflour. *J. Food Sci. Technol.* **2014**, *51*, 2600–2607. [CrossRef]
28. Codină, G.G.; Bordei, D.; Pâslaru, V. The effects of different doses of gluten on rheological behavior of dough and bread quality. *Rom. Biotechnol. Lett.* **2008**, *13*, 37–42.
29. Rojas, J.A.; Rosell, C.M.; Benedito de Barber, C. Pasting properties of different wheat flour-hydrocolloid systems. *Food Hydrocoll.* **1999**, *13*, 27–33. [CrossRef]
30. Qiu, S.; Yadav, M.P.; Chen, H.; Liu, Y.; Tatsumi, E.; Yin, L. Effects of corn fiber gums (CFG) on the pasting and thermal behaviours of maize starch. *Carbohydr. Polym.* **2015**, *115*, 246–252. [CrossRef]
31. Meng, Y.; Guan, X.; Liu, X.; Zhang, H. The rheology and microstructure of composite wheat dough enriched with extruded mung bean flour. *LWT Food Sci. Technol.* **2019**, *109*, 378–386. [CrossRef]
32. Zheng, M.; Su, H.; You, Q.; Zeng, S.; Zheng, B.; Zhang, Y.; Zeng, H. An insight into the retrogradation behaviors and molecular structures of lotus seed starch-hydrocolloid blends. *Food Chem.* **2019**, *295*, 548–555. [CrossRef] [PubMed]
33. Pareyt, B.; Finnie, S.M.; Putseys, J.A.; Delcour, J.A. Lipids in bread making: Sources, interactions, and impact on bread quality. *J. Cereal Sci.* **2011**, *54*, 266–279. [CrossRef]
34. Mironeasa, S.; Iuga, M.; Zaharia, D.; Mironeasa, C. Rheological analysis of wheat flour dough as influenced by grape peels of different particle sizes and addition levels. *Food Bioprocess Technol.* **2019**, *12*, 228–245. [CrossRef]
35. Yuksel, F.; Campanella, O.H. Textural, rheological and pasting properties of dough enriched with einkorn, cranberry bean and potato flours, using simplex lattice mixture design. *Qual. Assur. Saf. Crops Foods* **2018**, *10*, 389–398. [CrossRef]
36. Edel, A.L.; Aliani, M.; Pierce, G.N. Stability of bioactives in flaxseed and flaxseed-fortified foods. *Food Res. Int.* **2015**, *77*, 140–155. [CrossRef]

© 2019 by the authors. Licensee MDPI, Basel, Switzerland. This article is an open access article distributed under the terms and conditions of the Creative Commons Attribution (CC BY) license (http://creativecommons.org/licenses/by/4.0/).

Article

Whole Grain Muffin Acceptance by Young Adults

Thomas Mellette [1], Kathryn Yerxa [2,3], Mona Therrien [3] and Mary Ellen Camire [3,*]

1. WakeMed Health & Hospitals, Raleigh, NC 27610, USA; tommellette@gmail.com
2. Cooperative Extension, University of Maine, Orono, ME 04469-57417, USA; kate.yerxa@maine.edu
3. School of Food & Agriculture, University of Maine, Orono, ME 04469-5735, USA; mona.therrien@maine.edu
* Correspondence: camire@maine.edu; Tel.: +1-207-581-1627

Received: 30 April 2018; Accepted: 10 June 2018; Published: 13 June 2018

Abstract: Adolescents and young adults in the United States do not consume recommended amounts of whole grains. University dining services have opportunities to inform students about whole grains and to offer foods containing blends of whole grains with refined flour to increase daily consumption of these healthful foods. An online survey of university students (n = 100) found that 70% of respondents did not know the proportion of servings of whole grains that should be eaten daily. Mini blueberry muffins containing 50, 75, and 100% white whole wheat flour were served to 50 undergraduate students who rated their liking of the muffins using a nine-point hedonic scale. Respondents liked all muffin formulations similarly for appearance, taste, texture and overall liking. After the whole grain content of each muffin was revealed, 66% of students increased their liking of the muffins containing 100% whole wheat flour. Only half of the students increased their liking for the 75% whole wheat flour muffins, and most students reported no change in liking for the muffins made with the lowest percentage of whole wheat flour. Labeling whole grain foods in university foodservice operations may increase consumption of this food group by some students. Further research with actual purchase behavior is needed.

Keywords: whole grains; nutrition knowledge; consumer; baking; sensory evaluation

---

1. Introduction

Increased consumption of refined grains and lower consumption of whole grains has been associated with increased risks for developing health problems including obesity, cardiovascular disease, type 2 diabetes, and cancer [1–4]. Health Canada [5] and the United States Dietary Guidelines for Americans [6] recommend that half of all grain servings be whole grains. Foods made with enriched refined flour can contribute essential nutrients and comprise a significant portion of many Americans' diets according to the National Health and Nutrition Examination Survey (NHANES) [7]. Whole grain and dietary fiber consumption are low, thus Kranz et al. [8] recommended that public health messages focus on high fiber whole grain foods. Foods consumed at breakfast supplied more than 40% of whole grains consumed by children and adults according to the 2001–2010 NHANES, yet less than 20% of dietary fiber consumption occurred at breakfast in children and adults under the age of 51 years [9].

The eating habits of young adults are a concern since avoidance of whole grain foods leads to increased risk of disease later in life, and poor dietary choices may be passed on to the next generation [10]. Consumption of whole grains by 9–12-year-old children was dependent upon the availability of such foods at home [11]. Parents of young children in Northern Ireland became interested in introducing additional whole grain foods to their children after learning about the health benefits associated with these foods [12]. Larson et al. [13] concluded that for both sexes, availability of whole grain bread at home, penchant for the taste of whole grain bread, and self-efficacy to eat three or more servings of whole grains daily were strong (p < 0.001) predictors of whole grain consumption and frequency of fast food consumption was an indicator for lower whole grain intakes.

A diet recall study of 202 undergraduate college students found that 86% reported eating whole grains, but approximately 69% of all surveyed students did not consume the recommended three servings per day [14]. Rose et al. [15] reported that among 159 college students with a mean age of 19.9 years, fewer servings were consumed than by any other age group. Based on food records in that study, the college students consumed on average 0.7 servings of whole grains per day, which is 37.5% less than the 1.12 servings reported for all Americans [16].

One reason for this low whole grain intake could be due to a misunderstanding by college students of the dietary recommendations for whole grain consumption and what constitutes a whole grain food. A survey among 72 college students 18–23 years of age found that only 3% were able to identify the current Dietary Guidelines for Americans whole grain recommendations [17]. The majority of those survey respondents could not identify whether or not foods were made with whole grains. Undergraduate college students who had taken a nutrition course were more likely to know the current whole grain recommendations and were more likely to associate whole grain consumption with better health outcomes [18]. Magalis et al. [19] studied 69 California college students and reported that the students did not recognize or understand whole grains and over-estimated their whole grain consumption. College students 18–24 years old increased their whole grain consumption from 0.37 to 1.16 servings per day after taking a general nutrition course [20]. There is little information regarding the availability of whole grain foods on college campuses in the U.S. Whole grain intake among college students is partly dependent upon the availability of whole grain foods on campus [15]. An assessment of 15 college campuses across the U.S. between 2009 and 2011 for healthful dining options found that dining halls were more likely to have healthful food options than student union or snack-bar type locations, but the authors concluded that college dining locations overall offer limited healthy options [21].

Taste is among the top determinants of food choice for people in the U.S. [22]. Australian university students (n = 1306) who rated taste as highly important for food selection had poorer quality diets and were more likely to consume foods made with refined grains such as cakes, pastries, biscuits, and pizza [23]. Adolescents in the United Kingdom were aware that whole grain foods were more healthful than foods made with refined grains, but disliking of the sensory properties of whole grains was the primary obstacle to consumption [24]. For young adults, the association of poor taste with whole grains may create a bias against food products that have their whole grain content labeled prominently on the front of the package. A qualitative study of Northern Ireland adults' attitudes indicated that serving whole grains "in disguise" might counter the prevalent belief that whole grains have inferior sensory quality [12].

The main objectives of this research project were to understand the reasoning behind college students' whole grain food choices and whether students would accept muffins made with white whole wheat flour. The primary hypotheses for the online survey were that University of Maine students did not understand whole grain consumption recommendations and that believing whole grain foods were healthful would not influence their interest in consuming more whole grains. The research hypothesis for the acceptability study was that university students would find muffins made with white whole wheat flour to be acceptable and that liking would increase when the whole grain content was displayed.

2. Materials and Methods

2.1. Student Sample

The University of Maine is a land grant institution located in the greater Bangor region that has a population of approximately 153,000 people [25]. In the academic year 2014–2015, the university had an enrollment of 9339 undergraduates and 1947 graduate students. Among the undergraduate population, 48% were female, 78% were Caucasian, and 74% were Maine residents [26]. Students were recruited through the campus email conferencing system. Criteria for inclusion in the study were being aged 18–24 years and an undergraduate student at the University of Maine. Students majoring in food science or human nutrition were excluded from the study because they may have more knowledge about whole grains than would other students [18]; nutrition majors were more likely to follow the Dietary Guidelines for Americans to choose grains, fruits, and vegetables than were students in other majors [27]. Students in food science and nutrition represented less than 2% of the total undergraduates at the University of Maine matriculating at the time of the study. Persons with an allergy, intolerance, or aversion to foods containing wheat or dairy were asked to not participate in the sensory evaluation. The University of Maine's Institutional Review Board (IRB) approved the study protocols on March 7 2014 and judged the research exempt from further review.

2.2. Survey Instrument

A 20-question Internet survey was developed with Qualtrics software (v. 60262, Qualtrics LLC, Provo, UT, USA) was pilot-tested with 21 University of Maine faculty and students over the age of 18 years. The questions were designed to assess participants' access to and knowledge of whole grains and how well they liked whole grain products. Convenience sampling was used to recruit potential participants. When the interested parties selected the email link listed in the recruitment notice, they were brought to a web page showing the informed consent form. Prospective participants were informed of the $5.00 cash compensation for survey participation. Individuals could choose to either begin the survey or exit the window and not participate. At the end of the survey, participants were given the option of entering their e-mail address should they wish to receive further instruction on how to collect their compensation. The survey closed when 100 surveys were completed.

2.3. Sensory Evaluation

A wild blueberry muffin recipe [28] was modified to yield three treatments containing 50%, 75%, and 100% white whole wheat flour (King Arthur, Norwich, VT, USA) in combination with all-purpose flour (Hannaford Bros. Co., Scarborough, ME, USA) (Table 1). The 3 × 4 mini-muffin pans (30 mL volume; Wearever, Lancaster, OH, USA) were placed on a full-size sheet tray (The Vollrath Co., L.L.C., Sheboygan, WI, USA) and coated evenly with pan release spray (Par-Way Tryson Co. St. Clair, MO, USA). Seventeen g of muffin batter was then placed into each of the pan slots with a 22-mL scoop (The Vollrath Co., LLC, Sheboygan, WI, USA). The sheet tray was placed into the oven at 204 °C for a total of 30 min, with pan rotation every 10 min. The muffins were cooled on wire racks for 20 min and were placed into 7.1L plastic storage trays (Carlisle Companies, Inc., Charlotte, NC, USA). Each layer of muffins was separated with deli paper (James River Corp., Parchment, MI, USA) and the container was covered with plastic wrap (Reynolds Consumer Products LLC, Lake Forest, IL, USA) overnight before sensory evaluation.

After participants had read the informed consent form for sensory evaluation, they were escorted to one of 12 partitioned cubicles with climate control and simulated Northern daylight that meets ASTM guidelines for sensory evaluation laboratories [29]. Each cubicle contained a Microsoft® Windows computer. SIMS2000 sensory evaluation software (version 6.0, Sensory Computer Systems LLC, Berkeley Heights, NJ, USA) was used to create the questionnaire, randomize sample order presentation, and collect and analyze data. Demographic questions were asked first, and then participants received a tray containing one muffin for each of the three test formulations. Each formulation was labeled with a randomly-selected three-digit code. Sample order presentation was also randomized. All muffin samples were evaluated using a nine-point hedonic scale (1 = dislike extremely; 9 = like extremely) for appearance, flavor, texture, and overall liking [30]. Participants were also given spring water (DZA Brands LLC, Salisbury, NC, USA) to drink between each sample. After completing the questions for each randomized sample, the whole grain content and the number of servings of whole grains that a typically-sized muffin would provide were revealed to each participant. Participants were then asked whether overall liking for the sample was changed, if at all, by learning the amount of whole wheat flour present using a five-point Likert scale (decreased considerably, decreased somewhat, did not change, increased somewhat, increased considerably). Upon completion of the test, participants were given $10 cash compensation.

Table 1. Muffin formulations (g) [a,b].

Title 1	Manufacturer	% White Whole Wheat Flour		
		50%	75%	100%
white whole wheat flour	King Arthur Flour Co. Inc., Norwich, VT	235	352.5	470
all-purpose flour	Hannaford Bros. Co., Scarborough, ME	235	117.5	0
buttermilk	H.P. Hood LLC, Lynnfield, MA	625	625	625
light brown sugar	Hannaford Bros. Co., Scarborough, ME	355	355	355
frozen wild blueberries	Jasper Wyman & Son, Milbridge, ME	215	215	215
vegetable oil	Hannaford Bros. Co., Scarborough, ME	110	110	110
baking powder	Clabber Girl Corp., Terre Haute, IN	12.5	12.5	12.5
iodized salt	Morton, Chicago, IL	10	10	10
vanilla extract	ACH Food Co., San Francisco, CA	10	10	10
Sugar in the Raw®	Cumberland Packing Corp., Brooklyn, NY	10	10	10
baking soda	Hannaford Bros. Co., Scarborough, ME	5	5	5
cinnamon	McCormick Corp., Sparks, MD	5	5	5

[a] Recipe modified from King Arthur Flour Company, Inc. [21]; [b] total batter weight: 1827.5 g, producing 107 servings of 17 g muffins.

2.4. Color Analysis

Muffins without blueberries were baked for color measurement to assess the color of the muffin batter only since the number and size of berries per muffin could not be controlled. CIE $L^* a^* b^*$ color values were measured with a LabScan XE Spectrophotometer (Hunter Associates Laboratories, Inc., Reston, VA, USA) using Hunter Lab universal software (v. 4.10, Hunter Associates Laboratories, Inc., Reston, VA, USA). A port size of 50.8 mm and area view of 44.5 mm was used. Samples were placed on a watch glass (Corning Inc., Corning, NY, USA). Nine samples of each muffin variety were used to measure $L^* a^* b^*$ values. Muffins were cut in half vertically. The average of three readings for each of the nine samples was used for statistical analysis.

2.5. Statistical Analyses

Descriptive statistics for the survey were completed by the Qualtrics v. 60262 (Provo, UT, USA). Sensory evaluation data were analyzed with SYSTAT 12 software (v. 12.00.08, Systat Software, San Jose, CA, USA). Hedonic scores and changes in liking were compared by analysis of variance and Tukey's honestly significant difference test with a significance level of $p \leq 0.05$.

3. Results

3.1. Survey Results

3.1.1. Survey Respondent Characteristics

One hundred students completed the survey; the respondent gender distribution was 53% male and 46% female; one participant chose not to answer (Table 2). The University's undergraduate population at the time of the study was 52% male and 48% female [26]. The median age of the study participants and full-time students at the University was 20 years; the median year of school for survey respondents was second. Most respondents (53%) stated they lived on campus or in a sorority/fraternity building. The median age of those respondents who lived on campus was 19 years, and for those who lived off campus, 21 years. The number of participants on a college dining plan was split with 47% responding no and 47% responding yes, for most meals. Of those students who lived off campus, 80% reported that they somewhat or very much felt they had better access to whole grains than they would if they lived on campus, compared with 41% of students who lived on campus who claimed they had better access to whole grains than they would if they lived off campus.

Table 2. Demographic characteristics of University of Maine survey respondents.

Demographic Characteristic	Variable	Number of Respondents	Total Respondents
age (years)	18	14	90
	19	26	
	20	20	
	21	14	
	22	13	
	23	3	
gender	Female	46	99
	Male	53	
years in college	1	28	100
	2	29	
	3	25	
	4	14	
	5 or more	4	
housing situation	On campus or in a fraternity or sorority	53	100
	Off-campus	47	
dining plan	Yes	47	100
	No	47	
	Some meals	6	
weekly breakfast frequency	0–1	4	98
	2–3	23	
	4–6	32	
	7	39	

3.1.2. Food Habits of Survey Respondents

While daily breakfast was not a habit for 60% of respondents, the median days per week breakfast was eaten was six days. Twenty-seven students reported eating breakfast four or fewer days per week. Eighteen-year-olds were less likely to eat breakfast than were 20, 21, and 22-year-olds ($p \leq 0.05$). Students in their first year of university were less likely to eat breakfast than were third and fourth-year students ($p \leq 0.05$). There were no significant differences between gender and likelihood to eat breakfast. Those who lived on campus were less likely to eat breakfast than those who lived off campus ($p \leq 0.01$).

Participants were asked to rank four factors in order of importance when selecting a meal (data not shown). Ninety-three people answered this question. Taste was most often selected as being most important (n = 31), followed by health and cost with the same number of responses (n = 28), and convenience had the least amount of responses (n = 6). However, when the rank sums of all responses were compared using the Wilcoxon signed rank test, cost and taste were ranked equally important, health was less important than cost, convenience was ranked as least important.

To assess whether a 'health food' assumption existed towards whole grains, we asked participants if they considered whole grains to be a 'health food' (Table 3). All of the respondents answered this question, and 63 answered 'yes' while 25 said 'no' and the remaining 12 were 'not sure'. When asked how they agreed with the statement "health foods usually taste bad", 70% stated they either disagreed or strongly disagreed. Thirty-five percent of respondents reported that they either sometimes or always feel in general that whole grain foods would not taste good.

Table 3. University students' knowledge of and attitudes about whole grains.

Question	Variable	Number of Responses
Does whole grain on the label affect your purchase decision?	Not likely	10
	Neutral	35
	Likely	55
Do you in general feel that products made with whole grains will not taste good?	No	60
	Not Sure	4
	Sometimes	28
	Yes	7
Are whole grains a health food?	No	25
	Not Sure	12
	Yes	63
Do you prefer whole grains?	No	16
	No preference	30
	Yes	54
Reason for not choosing whole grain foods	Won't taste good	12
	Too expensive	38
	Too hard to prepare	7
	I always try to eat whole grains	26
	Does not apply to me	11
Avoiding gluten?	No	83
	Not Sure	3
	Sometimes	9
	Yes	5
Does avoiding gluten reduce your whole grain consumption?	No	2
	No Difference	11
	Yes	1

Respondents were asked to self-report their understanding of what whole grain means. Of those who answered (n = 99), 69% stated they felt that they had a good understanding of what whole grain means, and 31% either selected no or not sure. When asked what percentage of their total grain intake should be whole grain, 70% of respondents did not reply with the correct answer of 50%. In response to the question "do you in general prefer whole grain foods over non-whole grain foods?", 46% had no preference or answered no. Fifty-four percent reported that they prefer whole grain foods over non-whole grain foods. Females were more likely to prefer whole grain foods than males ($p \leq 0.01$). "Too expensive" was the top reason for not eating whole grains, followed by "it won't taste good".

Nearly 45% of survey respondents were either neutral to or not likely to purchase a packaged food with the words 'whole grain' clearly labeled on the front, and only 22% were very likely to buy such products. Females were more likely than males to purchase a packaged food with the words "whole grain" clearly labeled ($p \leq 0.01$). The gluten-free diet has become increasingly popular in American society both for those who require it as medical nutrition therapy and for those who avoid gluten by choice. Fourteen participants reported that they at least occasionally avoided eating foods that contain gluten, and all of these students avoided gluten-containing foods due to personal choice only. One respondent did state he/she ate less whole grains because of omitting gluten, two reported eating more whole grains, and 11 reported there is no difference in whole grain intake.

3.2. Consumer Evaluation of Whole Grain Muffins

3.2.1. Participant Demographic Traits

The median age of the 50 participants was 20 years, and the median year of university was the third year. The gender distribution was split evenly; two persons chose to not answer the question.

Although 44% of participants lived on campus or in a fraternity/sorority building, the majority of students were responsible for procuring their own food off-campus. Sixty-eight percent reported having a good understanding of what whole grain means, but 78% answered incorrectly when asked which percentage of total grain intake is recommended to be whole grain.

3.2.2. Hedonic Assessments

Mini-muffins were prepared so that a whole muffin could be served without creating sensory fatigue. The mean scores for the three samples of mini-muffins were between 6.5 and 7.2 (like slightly to like moderately) but no significant differences were found (Table 4). After each sample was evaluated, the percentage of white whole wheat flour used in each sample was revealed, and participants were asked to rate how their overall liking of the sample had changed, if at all, using a five-point scale: (1 = "decreased considerably," 3 = "no change," and 5 = "increased considerably"). The increases seen in overall liking of the 100% and 75% white whole wheat mini-muffins were significantly higher than those for the 50% whole wheat sample, and the 100% was re-rated higher than was the 75% whole wheat sample ($p \leq 0.05$) (Figure 1).

Table 4. Hedonic scores for blueberry muffins containing different percentages of white whole wheat flour [a,b].

Attribute	Percentage White Whole Wheat Flour		
	50%	75%	100%
appearance	7.1 ± 1.3	7.1 ± 1.6	7.0 ± 1.2
flavor	7.2 ± 1.4	7.1 ± 1.3	6.9 ± 1.5
texture	6.9 ± 1.4	6.7 ± 1.5	6.5 ± 1.6
overall liking	7.1 ± 1.3	7.1 ± 1.3	6.7 ± 1.5

[a] $n = 50$; $p > 0.05$, Tukey's honestly significant difference test. [b] 1 = dislike extremely; 5 = neither like nor dislike; 9 = like extremely. $n = 50$. Standard deviations are shown in parentheses next to means.

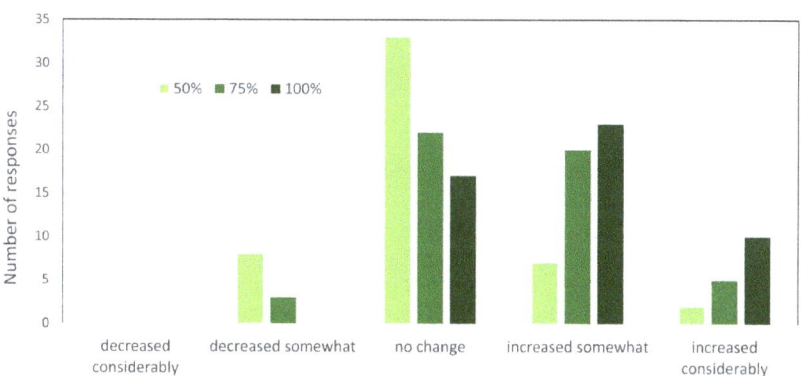

Figure 1. Change in overall acceptance of muffins after the revelation of whole grain content [a]. [a] Muffins contained 50–100% whole wheat flour.

3.3. Color

Since it is difficult to control the amount of blueberries per muffin and any associated anthocyanin bleeding, muffins were prepared without blueberries solely for crumb color analysis. The exterior CIE L^* and b^* of the muffins did not vary among the treatments; CIE a^* was significantly ($p \leq 0.001$) less red for the 50% whole muffins. Although the interior crumb color was significantly different between the 100% whole wheat and the other muffin types (Table 5), the differences were relatively small.

Table 5. CIE interior color of muffins containing different percentages of white whole wheat flour [A,B].

Whole Wheat Flour (%)	L^*	a^*	b^*
50	50.4 ± 1.3 a	8.9 ± 0.2 c	27.4 ± 0.8 b
75	48.1 ± 1.1 b	9.3 ± 0.2 b	27.4 ± 0.2 b
100	46.5 ± 1.5 c	9.8 ± 0.2 a	28.1 ± 0.5 a

[A] Means ± standard deviations (n = 9). Color scales: L^*, 0 = black, 100 = white; a^*, +a = red, −a = green; b^*, +b = yellow, −b = blue. [B] Values with the same letter in columns are not significantly different (p > 0.05).

4. Discussion

4.1. Survey Findings

The demographics of this study's participants were representative of the University's undergraduate population at the time of the study. Human nutrition/dietetics and food science students were excluded from the study because persons with nutritional education are more likely to know about whole grains and are thus more likely consume whole grain products [20,27]. The median age and year of school were very similar to other comparable whole grain studies involving college students [14,31]. Two- and four-year college students in Minneapolis (n = 1013) reported eating breakfast only 4.2 days per week [32]. Approximately 25% of participants in our survey reported eating breakfast four or fewer days per week. These are critical findings because Americans eat the majority of their whole grain intake during breakfast [16]. The present survey's findings suggested that students became more inclined to eat breakfast during their later college years, but this study lacked statistical power to support this inference. Students who lived off campus (80%) felt that they had better access to whole grain foods than they would if they lived on campus. Thirty-five percent of those who lived on campus felt that they would have better access to whole grains if they lived off campus. While healthful options may be limited at colleges across the country [21], data were not found on the availability of whole grain options in campus dining facilities. Increased availability of whole grains in restaurants as well as homes has been advocated to increase whole grain consumption by adolescents and young adults [13]. University dining services often provide multiple meals per day to undergraduate students and therefore present an opportunity to expose students to these foods.

When survey participants ranked their top four reasons for food choices, taste was most often selected as being number one, which is consistent with other research in the United States [22,33]. Whole grains are perceived to cost more than refined grain products [12]. The perceived healthfulness of whole grains and the notion that 'health foods' are likely to have poor sensory characteristics were explored, and findings were consistent with previous research that studied the reasons why people choose not to eat whole grains. One study showed that of those who reported they did not eat whole grain products, 45% indicated this was because they disliked the taste or texture [14]. A focus group study among adults found that many of the participants disliked whole grain products because of their sensory characteristics [12]. While taste is certainly relative to the individual, this may indicate that further education is needed to demonstrate that whole grain foods can be prepared in a variety of favorable methods.

To measure knowledge of whole grains and whole grain recommendations, participants were asked to self-report their perceived knowledge level. However, less than a third of respondents in this study who answered that they had a good understanding could correctly identify which percentage of daily grains should be whole grains. This finding indicates the need for nutrition education among these individuals. In a study comparing college students who have had a nutrition course and those who have not, those who had taken a nutrition course were both more likely to know whole grain recommendations and more likely to correctly identify whole grain products [14]. Given the increasing prevalence of chronic diseases that can be prevented with consumption of whole grains, a general nutrition course should be encouraged for all college majors. At the University of Maine, for example, the introductory food and nutrition class satisfies the general education requirement for applications of scientific knowledge.

Females were more likely to prefer whole grain foods and were also more likely to purchase a packaged food with the words 'whole grain' clearly labeled on the front of the package ($p \leq 0.05$). Nearly half ($n = 45$) of survey respondents were either neutral or not likely to purchase a packaged food with the words 'whole grain' clearly labeled on the front. This result further strengthens the hypothesis that for those with a bias against whole grain foods, whole grain labeling may deter students from purchasing those foods. However, for those who do prefer whole grain products, clear package labeling appears to be helpful for whole grain identification. People who seek out whole grain products often rely on food packaging or advertisements for guidance [18,20]. Therefore, it may be difficult to market a single whole grain product to both those who prefer whole grains and those who do not.

To overcome this challenge, the substitution of white whole wheat flour for regular all-purpose flour in various foods may be advantageous along with not labeling whole grain content on the front of the package. The students who prefer whole grain foods may read the ingredients list and identify that the product is made with whole grains and those who do not prefer whole grains may find sensory characteristics of the product appealing. White whole wheat tends to have a mild flavor, smooth texture, and a light color that is aesthetically pleasing to many consumers [34]. The sensory evaluation component of this study tested the hypothesis that a product made with white whole wheat flour that does not have its whole grain content clearly labeled is acceptable to college students. Undergraduate students rated white whole wheat muffins as more healthy ($p \leq 0.05$) after researchers revealed that the muffins contained whole wheat [35].

Among the various reported reasons for not eating whole grain foods, "too expensive" was the top response, followed by "it won't taste good" and "too hard to prepare". Taste, expense, and convenience were the top three barriers to whole grain consumption in another study [22]. Among 545 grain-based products evaluated, those containing the whole grain stamp typically cost four cents more per serving of the food [36]. It was not clear how much of the higher cost could be attributed to the fee charged to companies by the Whole Grains Council for the right to display the stamp.

Gluten-free diets have become increasingly popular, but people following such a diet may not have a medical condition that requires avoidance of gluten [37]. Participants in the nurses' health study and the health professionals' follow-up study who avoided gluten were more likely to have lower consumption of whole grains, and thus higher risk for cardiovascular disease [38]. Fourteen of the survey respondents in this study reported sometimes avoiding eating foods that contain gluten, and all of these students reported they avoided gluten by personal choice, not due to a confirmed medical diagnosis of celiac disease or gluten intolerance. There may be an opportunity for qualified health professionals such as registered dietitian nutritionists (RDN) to use the preferred media of television and the internet along with in-person education opportunities to distribute evidence-based nutrition information to college students. Education about whole grain gluten-free foods may also be needed for persons with celiac disease. In 2006, Cousineau et al. [39] reported that a web-based personalized nutrition curriculum might be helpful in distributing nutrition information to college students. The growth in popularity of smartphones and other technologies suggest that there may be more opportunities to promote healthful behaviors, especially for students with an internal and powerful other locus of control [40].

4.2. Sensory Evaluation

Muffins are a popular breakfast food and thus were selected as the test food in this study despite the added sugar in the formulation. A U.S. standard-sized muffin weighs 139 g [41]. Standard muffins made according to the formulations that we used would provide 18, 27, and 36 g of whole grain for the 50, 75, and 100% whole wheat flour formulations, respectively. No significant differences in hedonic ratings were found for any of the tested attributes (appearance, flavor, texture, overall liking). Muffins made with only whole wheat were liked as well as those made with mixtures of refined and whole wheat flours. Even when whole grain content was not prominently labeled, the college students in this study found the product acceptable. However, this study reports on a small sample ($n = 50$) at a small university in a racially-homogenous state. Neo and Brownlee [42] assessed liking of some whole grain foods among young adults in Singapore aged 21–26 years old. A two-week familiarization period increased acceptance of whole grain cookies, granola bars, muesli, and whole grain pasta. The Willingness to Eat Whole Grains Questionnaire has been recently evaluated with university students aged 18–29 years and could be a useful tool in future nutrition education and intervention projects [43].

4.3. Strengths and Limitations

The survey yielded information to better inform university dining services about students interests and potential barriers to acceptance of whole grain foods. Limitations of the survey included incomplete representation of the student population and lack of self-reported whole grain consumption data and assessment of students' ability to recognize whole grain foods. The whole grain muffin sensory evaluation provided interesting information about university students' willingness to like the muffins containing a greater percentage of whole wheat flour, but we cannot rule out the possibility that students sought to please the researchers. Drawbacks of the experimental design were the lack of a control sample containing no whole wheat, and the inclusion of only white whole wheat flour. Comparisons of products made with refined flour, white whole wheat, and red whole wheat are needed. Baking conditions were optimized for muffins containing whole wheat, thus potentially causing an inferior muffin containing only refined flour. Future work should include bread, wraps, and other non-sweet types of grain products, and involve students from several universities since the University of Maine is a relatively small institution without a diverse student population. Attitudes towards and knowledge of whole grains by young adults who are not pursuing university degrees are other important topics that should be considered for future research.

Nutrition behavior research on healthful food selection has focused on children since interventions at an early age are expected to lead to lifelong habits. A systematic review of children's food choices identified four important learning processes: familiarization, observational learning, associative learning, and categorization of foods [44]. A greater understanding of the processes used by young adults to select foods as they transition to independent living is needed. Familiarization with whole grain foods was evaluated in a group of overweight adults who consumed less than 20 g of whole grains daily [45]. The intervention groups were provided with whole grain foods and asked to consume three to six servings per day for 16 weeks. Food frequency questionnaires collected over the next 12 months indicated that persons in the intervention groups consumed more whole grains than did the control group members immediately after the intervention period ended, but whole grain consumption declined with time and never approached the 60–120 g levels advocated during the intervention period. Universities have the opportunity to expose students to whole grain foods for periods of approximately eight months per year, thus longer interventions with university students should be conducted.

5. Conclusions

Universities have the opportunity to educate students about whole grain foods and to develop healthful eating patterns that include whole grains. The inclusion of more whole grain foods at breakfast in university dining halls may reach those students who do eat breakfast, but younger students are more likely to skip this meal. Identification of whole grains at local restaurants and markets might assist students as they begin expanding their food options and preparing their meals when they move off-campus. Gender-specific nutrition education messages might be effective [29,46]. Establishment of regular whole grain consumption patterns in young adults could lead to increased popularity of these healthful foods in subsequent generations. White wheat offers foodservice operations and food processors an option to make whole grains more appealing to young adults, particularly those enrolled in a university.

Author Contributions: T.M. was responsible for conceptualization, formal analysis, and methodology. T.M., K.Y., and M.T. reviewed and edited the draft manuscript. M.E.C. was responsible for project administration and writing the original draft. All authors read and approved the final manuscript.

Funding: Maine Agricultural and Forest Experiments Station Publication Number #3608. This project was supported by the USDA National Institute of Food and Agriculture, Hatch Project Number ME021804 through the Maine Agricultural & Forest Experiment Station.

Acknowledgments: Kathleen Crosby, Michael Dougherty, Derek Ford, and Sierra Guay provided assistance with sensory evaluation.

Conflicts of Interest: The authors declare that they have no conflicts of interest.

References

1. Thielecke, F.; Jonnalagadda, S.S. Can whole grain help in weight management? *J. Clin. Gastroenterol.* **2014**, *48* (Suppl. 1), S70–S77. [CrossRef] [PubMed]
2. Aune, D.; Keum, N.; Giovannucci, E.; Fadnes, L.T.; Boffetta, P.; Greenwood, D.C.; Tonstad, S.; Vatten, L.J.; Riboli, E.; Norat, T. Whole grain consumption and risk of cardiovascular disease, cancer, and all cause and cause specific mortality: Systematic review and dose-response meta-analysis of prospective studies. *BMJ* **2016**, *353*, i2716. [CrossRef] [PubMed]
3. Bellou, V.; Belbasis, L.; Tzoulaki, I.; Evangelou, E. Risk factors for type 2 diabetes mellitus: An exposure-wide umbrella review of meta-analyses. *PLoS ONE* **2018**, *13*, e0194127. [CrossRef] [PubMed]
4. Li, B.; Zhang, G.; Tan, M.; Zhao, L.; Jin, L.; Tang, X.; Jiang, G.; Zhong, K. Consumption of whole grains in relation to mortality from all causes, cardiovascular disease, and diabetes: Dose-response meta-analysis of prospective cohort studies. *Medicine* **2016**, *95*, e4229. [CrossRef] [PubMed]
5. Health Canada. Eating Well with Canada's Food Guide. Available online: https://www.canada.ca/content/dam/hc-sc/migration/hc-sc/fn-an/alt_formats/hpfb-dgpsa/pdf/food-guide-aliment/view_eatwell_vue_bienmang-eng.pdf (accessed on 7 April 2018).
6. U.S. Department Health Human Services; United States Department of Agriculture. *2015–2020 Dietary Guidelines for Americans*, 8th ed. Available online: https://health.gov/dietaryguidelines/2015/guidelines/ (accessed on 11 June 2018).
7. Papanikolaou, Y.; Fulgoni, V. Grain foods are contributors of nutrient density for American adults and help close nutrient recommendation gaps: Data from the National Health and Nutrition Examination Survey, 2009–2012. *Nutrients* **2017**, *9*, 873. [CrossRef] [PubMed]
8. Kranz, S.; Dodd, K.; Juan, W.; Johnson, L.; Jahns, L. Whole grains contribute only a small proportion of dietary fiber to the U.S. Diet. *Nutrients* **2017**, *9*, 153. [CrossRef] [PubMed]
9. McGill, C.R.; Fulgoni, V.L.; Devareddy, L. Ten-year trends in fiber and whole grain intakes and food sources for the United States population: National Health and Nutrition Examination Survey 2001–2010. *Nutrients* **2015**, *7*, 1119–1130. [CrossRef] [PubMed]
10. Yee, A.Z.; Lwin, M.O.; Ho, S.S. The influence of parental practices on child promotive and preventive food consumption behaviors: A systematic review and meta-analysis. *Int. J. Behav. Nutr. Phys. Act.* **2017**, *14*, 47. [CrossRef] [PubMed]
11. Rosen, R.A.; Burgess-Champoux, T.L.; Marquart, L.; Reicks, M.M. Associations between whole-grain intake psychosocial variables, and home availability among elementary school children. *J. Nutr. Educ. Behav.* **2012**, *44*, 628–633. [CrossRef] [PubMed]
12. McMackin, E.; Dean, M.; Woodside, J.V.; McKinley, M.C. Whole grains and health: Attitudes to whole grains against a prevailing background of increased marketing and promotion. *Public Health Nutr.* **2013**, *16*, 743–751. [CrossRef] [PubMed]
13. Larson, N.I.; Neumark-Sztainer, D.; Story, M.; Burgess-Champoux, T. Whole-grain intake correlates among adolescents and young adults: Findings from Project EAT. *J. Am. Dietet. Assoc.* **2010**, *110*, 230–237. [CrossRef] [PubMed]

14. Bisanz, K.J.; Krogstrand, K.L. Consumption & attitudes about whole grain foods of UNL students who dine in a campus cafeteria. *Rev. Undergrad. Res. Agric. Life Sci.* **2007**, *2*, 1–16.
15. Rose, N.; Hosig, K.; Davy, B.; Serrano, E.; Davis, L. Whole-grain intake is associated with body mass index in college students. *J. Nutr. Educ. Behav.* **2007**, *39*, 90–94. [CrossRef] [PubMed]
16. Lin, B.-H.; Yen, S.T. *The U.S. Grain Consumption Landscape: Who Eats Grain, in What Form, Where, and How Much*; U.S. Department of Agriculture, Economic Research Service (USDA ERS): Washington, DC, USA, 2007.
17. Gager, E.; Agel, M.; Ling, H.; Vineis, M.; Policastro, P. Assessing college students' consumption and knowledge of whole grains, including identification and USDA recommendations. *J. Am. Dietet. Assoc.* **2011**, *111*, A79. [CrossRef]
18. Williams, B.A.; Mazier, M.J.P. Knowledge, perceptions, and consumption of whole grains among university students. *Can. J. Dietet. Pract. Res.* **2013**, *74*, 92–95. [CrossRef] [PubMed]
19. Magalis, R.M.; Giovanni, M.; Silliman, K. Whole grain foods: Is sensory liking related to knowledge, attitude, or intake? *Nutr. Food Sci.* **2016**, *46*, 488–503. [CrossRef]
20. Ha, E.J.; Caine-Bish, N. Interactive introductory nutrition course focusing on disease prevention increased whole-grain consumption by college students. *J. Nutr. Educ. Behav.* **2011**, *43*, 263–267. [CrossRef] [PubMed]
21. Horacek, T.M.; Erdman, M.B.; Byrd-Bredbenner, C.; Carey, G.; Colby, S.M.; Greene, G.W.; Guo, W.; Kattelmann, K.K.; Olfert, M.; Walsh, J.; et al. Assessment of the dining environment on and near the campuses of fifteen post-secondary institutions. *Public Health Nutr.* **2013**, *16*, 1186–1196. [CrossRef] [PubMed]
22. Blanck, H.M.; Yaroch, A.L.; Atienza, A.A.; Yi, S.L.; Zhang, J.; Masse, L.C. Factors influencing lunchtime food choices among working Americans. *Health Educ. Behav.* **2009**, *36*, 289–301. [CrossRef] [PubMed]
23. Kourouniotis, S.; Keast, R.S.J.; Riddell, L.J.; Lacy, K.; Thorpe, M.G.; Cicerale, S. The importance of taste on dietary choice, behaviour and intake in a group of young adults. *Appetite* **2016**, *103*, 1–7. [CrossRef] [PubMed]
24. Kamar, M.; Evans, C.; Hugh-Jones, S. Factors influencing adolescent whole grain intake: A theory-based qualitative study. *Appetite* **2016**, *101*, 125–133. [CrossRef] [PubMed]
25. Data USA Bangor, ME Metro Area. 2017. Available online: https://datausa.io/profile/geo/bangor-me-metro-area/ (accessed on 21 May 2018).
26. University of Maine Office of Institutional Research. Common Data Set 2014–2015. Available online: https://umaine.edu/oir/wp-content/uploads/sites/205/2015/04/CDS_2014-2015-4.30.15ls.pdf (accessed on 20 May 2018).
27. Hong, M.Y.; Shepanski, T.L.; Gaylis, J.B. Majoring in nutrition influences BMI of female college students. *J. Nutr. Sci.* **2016**, *5*, e8. [CrossRef] [PubMed]
28. Anonymous. 100% Whole Wheat Blueberry Muffins. Available online: http://www.kingarthurflour.com/recipes/100-whole-wheat-blueberry-muffins-recipe (accessed on 22 May 2014).
29. Kuesten, C.; Kruse, L. *Physical Requirement Guidelines for Sensory Evaluation Laboratories*, 2nd ed.; ASTM International: West Conshohocken, PA, USA, 2009; ISBN EB 978-0-8031-7074-2.
30. Peryam, D.R.; Pilgrim, F.J. Hedonic scale method of measuring food preference. *Food Technol.* **1957**, *11*, 9–14.
31. Matthews, J.I.; Doerr, L.; Dworatzek, P.D.N. University students intend to eat better but lack coping self-efficacy and knowledge of dietary recommendations. *J. Nutr. Educ. Behav.* **2016**, *48*, 12–19.e11. [CrossRef] [PubMed]
32. Laska, M.N.; Hearst, M.O.; Lust, K.; Lytle, L.A.; Story, M. How we eat what we eat: Identifying meal routines and practices most strongly associated with healthy and unhealthy dietary factors among young adults. *Public Health Nutr.* **2015**, *18*, 2135–2145. [CrossRef] [PubMed]
33. Glanz, K.; Basil, M.; Maibach, E.; Goldberg, J.; Snyder, D. Why Americans eat what they do: Taste, nutrition, cost, convenience, and weight control concerns as influences on food consumption. *J. Am. Dietet. Assoc.* **1998**, *98*, 1118–1126. [CrossRef]
34. Doblado-Maldonado, A.F.; Pike, O.A.; Sweley, J.C.; Rose, D.J. Key issues and challenges in whole wheat flour milling and storage. *J. Cereal Sci.* **2012**, *56*, 119–126. [CrossRef]
35. Camire, M.E.; Bolton, J.; Jordan, J.J.; Kelley, S.; Oberholtzer, A.; Qiu, X.; Dougherty, M.P. Color influences consumer opinions of wheat muffins. *Cereal Foods World* **2006**, *51*, 274–276. [CrossRef]
36. Mozaffarian, R.S.; Lee, R.M.; Kennedy, M.A.; Ludwig, D.S.; Mozaffarian, D.; Gortmaker, S.L. Identifying whole grain foods: A comparison of different approaches for selecting more healthful whole grain products. *Public Health Nutr.* **2013**, *16*, 2255–2264. [CrossRef] [PubMed]

37. Blackett, J.W.; Shamsunder, M.; Reilly, N.R.; Green, P.H.R.; Lebwohl, B. Characteristics and comorbidities of inpatients without celiac disease on a gluten-free diet. *Eur. J. Gastroenterol. Hepatol.* **2018**, *30*, 477–483. [CrossRef] [PubMed]
38. Lebwohl, B.; Cao, Y.; Zong, G.; Hu, F.B.; Green, P.H.R.; Neugut, A.I.; Rimm, E.B.; Sampson, L.; Dougherty, L.W.; Giovannucci, E.; et al. Long term gluten consumption in adults without celiac disease and risk of coronary heart disease: Prospective cohort study. *BMJ* **2017**, *357*, j1892. [CrossRef] [PubMed]
39. Cousineau, T.M.; Franko, D.L.; Ciccazzo, M.; Goldstein, M.; Rosenthal, E. Web-based nutrition education for college students: Is it feasible? *Eval. Program Plan.* **2006**, *29*, 23–33. [CrossRef] [PubMed]
40. Bennett, B.L.; Goldstein, C.M.; Gathright, E.C.; Hughes, J.W.; Latner, J.D. Internal health locus of control predicts willingness to track health behaviors online and with smartphone applications. *Psychol. Health Med.* **2017**, *22*, 1224–1229. [CrossRef] [PubMed]
41. United States Department of Agriculture Food Composition Databases. Available online: https://ndb.nal.usda.gov/ndb/search/list (accessed on 22 May 2018).
42. Neo, J.E.; Brownlee, I.A. Wholegrain food acceptance in young Singaporean adults. *Nutrients* **2017**, *9*, 371. [CrossRef] [PubMed]
43. Tuuri, G.; Cater, M.; Craft, B.; Bailey, A.; Miketinas, D. Exploratory and confirmatory factory analysis of the Willingness to Eat Whole Grains Questionnaire: A measure of young adults' attitudes toward consuming whole grain foods. *Appetite* **2016**, *105*, 460–467. [CrossRef] [PubMed]
44. Mura Paroche, M.; Caton, S.J.; Vereijken, C.M.J.L.; Weenen, H.; Houston-Price, C. How infants and young children learn about food: A systematic review. *Front. Psychol.* **2017**, *8*, 1046. [CrossRef] [PubMed]
45. Brownlee, I.A.; Kuznesof, S.A.; Moore, C.; Jebb, S.A.; Seal, C.J. The impact of a 16-week dietary intervention with prescribed amounts of whole-grain foods on subsequent, elective whole grain consumption. *Br. J. Nutr.* **2013**, *110*, 943–948. [CrossRef] [PubMed]
46. Boek, S.; Bianco-Simeral, S.; Chan, K.; Goto, K. Gender and race are significant determinants of students' food choices on a college campus. *J. Nutr. Educ. Behav.* **2012**, *44*, 372–378. [CrossRef] [PubMed]

© 2018 by the authors. Licensee MDPI, Basel, Switzerland. This article is an open access article distributed under the terms and conditions of the Creative Commons Attribution (CC BY) license (http://creativecommons.org/licenses/by/4.0/).

Article

Influence of Flour Particle Size Distribution on the Quality of Maize Gluten-Free Cookies

Mayara Belorio *, Marta Sahagún and Manuel Gómez

College of Agricultural Engineering, University of Valladolid, 34004 Palencia, Spain; msahaguncs@gmail.com (M.S.); pallares@iaf.uva.es (M.G.)
* Correspondence: beloriom@gmail.com; Tel.: +34-979-108-495

Received: 29 January 2019; Accepted: 20 February 2019; Published: 23 February 2019

Abstract: The objective of the present study was to analyse the influence of particle size distribution of maize flour in the formulation of gluten-free cookies. Different cookie formulations were made with three distinct maize flour fractions obtained by sieving (less than 80 μm; between 80 and 180 μm; greater than 180 μm). Cookies dimension, texture and colour were evaluated. Flour hydration properties and cookie dough rheology were also measured. Overall, an increase in maize flour particle size decreases the values of water holding capacity (WHC), swelling volume and G' (elastic modulus) for the doughs. An increase in average particle size also increases diameter and spread factor of the cookies but decreases their hardness. A higher percentage of thick particles is more effective to reduce cookie hardness, but a certain percentage of thinner particles is necessary to give cohesion to the dough and to allow formation of the cookies without breaking. Cookies with a larger diameter also presented a darker colour after baking.

Keywords: maize flour; gluten-free; cookie; particle size; sieve

1. Introduction

Coeliac disease is an autoimmune disease related to an intolerance of gluten, which affects adults and children. It is treated by restricting gluten-containing food in the diet [1], which significantly increases the demand for gluten-free products. Therefore, to take advantage of this growth, many companies are looking to diversify and develop new products that meet this demand.

Most processed and pre-packaged bakery products, such as breads, cakes and cookies, are commonly produced with wheat flour. Although gluten plays an important role in bakery food processes, a glutenous structure is not developed in most types of cookies because of the high fat and sugar content in recipes and the scarce mechanical work imparted in the mixing process [2,3]. Thus, it is possible to obtain gluten-free cookies with characteristics very similar to those made with wheat flour [3–5]. Amongst cookies made with gluten-free flours, those formulated with maize flour are rated the highest by consumers [3].

In the case of cookies made with wheat flour, some characteristics, such as protein content or water absorption capacity, can justify its use in cookie preparation [2]. Maize, rice or legume flours are usually coarser than wheat flour due to their harder grains. It is already known that the particle size of wheat flour can influence cookie quality [6] but it could also be true for gluten-free flours. To investigate this, Rao et al. [7] observed that coarse sorghum flours produced cookies with lower hardness and better consumer acceptability. Mancebo et al. [8] also reported that coarse rice flours produced cookies with higher spread factors and lower hardness, which agrees with the observations by Ai et al. [9] with bean powders. However, Mancebo et al. [8] did not observe this effect in maize cookies, since both types of maize flour they used exhibited minor differences in particle size.

In some research papers, different particle size is obtained by forcing the grinding process until particles with a finer size are obtained [9,10]. However, this mechanical process causes an increase in damage starch that also affects how cookies develop during elaboration after elaboration [6,11]. There are few studies that achieve distinct fractions by sieving [3,7], nor are there any studies into blends of different flour fractions with different distributions of particle size.

Thus, the present research proposes studying how different fractions of sieved white maize flour, both alone and in combination, could influence gluten-free cookie dough properties (hydration and rheology) and cookie quality (physical properties, texture and colour).

2. Materials and Methods

2.1. Cookie Ingredients

The white maize flour (5.87 g/100 g protein) used in this study was produced by Molendum Ingredients S.L. (Zamora, Spain). A Bühler MLI 300B sifter (Milan, Italy) holding 80 and 180 µm sieves was used for 15 minutes to obtain three maize flour fractions: A (<80 µm), B (80–180 µm) and C (>180 µm).

Other ingredients used in the cookie recipe were white sugar (AB Azucarera Iberia, Valladolid, Spain), margarine (Argenta crema, Puratos, Barcelona, Spain), sodium bicarbonate (Manuel Riesgo S. A., Madrid, Spain) and local tap water.

2.2. Flour Characterisation

Flour particle size was evaluated using a Mastersizer 3000 particle size analyser (Malvern Instruments, Malvern, UK). Values of D[4,3], which represents the equivalent spherical diameter of the particles, and of D(10), D(50) and D(90), which represent the maximum particle diameter below which 10%, 50% and 90% of the sample fall, respectively, were obtained. All the measurements were carried out in duplicate.

Regarding hydration, water binding capacity (WBC, i.e., the amount of water retained by the sample after it has been centrifuged) was measured as described in AACC International method 56-30.01 [12]. Water holding capacity (WHC, i.e., the amount of water retained by the sample without being subjected to any stress) and swelling volume (SV, i.e., the volume occupied by a known weight of sample) were measured as described by Mancebo et al. [3]. All hydration properties were analysed in duplicate.

2.3. Cookie Formulation

The original maize flour and the individual fractions were applied in different percentages giving rise to eleven combinations, as shown in Table 1. The mixing process was carried out as described by Mancebo et al. [8]. For the rheology test, the dough was rolled out to 3 mm thickness on a dough sheeter and then cut into round shape with a cutter of 60 mm diameter. For the baking test, samples were rolled out to 6 mm thickness and were cut with a 40 mm diameter cutter. The baking process was performed in a baking oven at 185 °C for 14 min. The cookies were cooled for one hour at room temperature and then placed in plastic bags, to avoid the interference of humidity. They were stored for seven days in a chamber with controlled temperature (25 °C) for further analysis. All cookie baking was done in duplicate.

Table 1. Gluten-free cookie formulations presented in grams with different percentages of maize flour and fractions A (<80 μm), B (80–180 μm) and C (>180 μm).

Ingredients	CF	100A	100B	100C	50A/50B	50A/50C	50B/50C	25A/75B	75A/25B	25A/75C	75A/25C
White maize flour	173.2	-	-	-	-	-	-	-	-	-	-
A (<80 μm)	-	173.2	-	-	86.6	86.6	-	43.3	129.9	43.3	129.9
B (80–180 μm)	-	-	173.2	-	86.6	-	86.6	129.9	43.3	-	-
C (>180 μm)	-	-	-	173.2	-	86.6	86.6	-	-	129.9	43.3
White sugar	124.8	124.8	124.8	124.8	124.8	124.8	124.8	124.8	124.8	124.8	124.8
Margarine	77.6	77.6	77.6	77.6	77.6	77.6	77.6	77.6	77.6	77.6	77.6
Sodium bicarbonate	3.6	3.6	3.6	3.6	3.6	3.6	3.6	3.6	3.6	3.6	3.6
Tap water	25.0	25.0	25.0	25.0	25.0	25.0	25.0	25.0	25.0	25.0	25.0

CF: control formulation.

2.4. Dough Rheology

Rheological behaviour of the fresh cookie dough was evaluated using a rheometer (Haake RheoStress 1, Thermo Fisher Scientific, Scheverte, Germany). Each dough sample was placed on a titanium parallel-serrated plate geometry PP60 Ti (60 mm diameter, 3 mm gap) and covered with Panreac Vaseline oil (Panreac Química S.A., Castellar del Vallés, Spain) to avoid drying during the test. A Phoenix II P1-C25P water bath maintained the temperature at 25 °C.

In the first measurement, the cookie dough was subjected to a strain sweep (stress range of 0.1–100 Pa) at a constant temperature (25 °C) and frequency (1 Hz) to identify the linear viscoelastic region. Then, using these results, a stress value within the linear viscoelastic region was selected and applied in a frequency sweep test to obtain the values of the elastic modulus (G', Pa), viscous modulus (G'', Pa), complex modulus (G^*) and tan delta (G''/G') over a range of frequency values (w, Hz). The measurements were made in duplicate.

2.5. Cookie Characteristics

Cookie moisture was evaluated as described by AACC method 44-15.02 [13] and the test was made in duplicate.

Cookie diameter was measured twice, in perpendicular directions, to achieve an average diameter (D). Cookie thickness (T) was also measured to obtain the spread factor (D/T).

Texture parameters of the cookies were analysed by using a Texture TA-XT2 texture analyser (Stable Micro Systems, Surrey, UK). The peak force, or hardness, (N) and the elastic modulus (N/mm^2) were obtained by the compression of a 'three-point bending' test with a three-point bending rig probe (HDP/3PB). The measurement conditions were: travel distance of 20 mm, trigger force of 5 g and test speed of 2.0 mm/s.

Cookie colour was measured at the centre of the surface crust with a Minolta CN-508i spectrophotometer (Minolta Co. Ltd., Osaka, Japan) using a D65 illuminant with a 2° standard observer angle. L^*, a^* and b^* values were expressed in the colour space defined by the International Commission on Illuminaion (CIE).

All cookie characteristics were measured in six cookies of each batch, seven days after baking.

2.6. Statistical Analysis

Statistical analysis of the differences between the parameters of the different formulations were evaluated by analysis of variance (ANOVA) using Statgraphics Centurion XVI software (StatPoint Technologies Inc., Warrenton, DC, USA). Fisher's least significant difference (LSD) was used to describe means with 95% confidence intervals.

3. Results

3.1. Particle Size and Hydration Properties

As shown in Table 2, the different blends presented an average size D[4,3] within the range of particle sizes of the fractions that comprised it. As expected, the average particle size of blends formed from fractions A and B increased with the quantity of fraction B. This also occurred with the mixtures of fractions A and C. The control sample had an average particle size slightly higher than fraction B, but clearly lower than fraction C. D(10) and D(50) values decreased as the percentage of fraction A increased. However, for mixtures containing 50% or more of A, there were no further significant changes, regardless of whether they were mixed with B or C. This is logical if it is remembered that these values refer to 10% or 50% of the thinnest flours from the mixtures obtained from fraction A. In the case of mixtures with 25% of A, there was a significant difference in D(50) values and they were even smaller in mixtures with the fraction B. Control flour had relatively low D(10), which was similar to mixtures with 25% of A and smaller than mixtures without fraction A. For its part, the D(50) is minor in all mixtures with A, with just one exception (25A/75C), and no significant differences were observed with fraction B. In general, the average sizes of the flours used in this research were similar to those applied in other studies on gluten-free cookie formulations [3,4,7].

Table 2. Maize flour, maize flour fractions and their combinations particle size measurements.

	D[4,3]	D(10)	D(50)	D(90)
Control flour	199.9 ± 6.4 g	26.5 ± 1.4 c	178.8 ± 6.5 e	401.4 ± 8.8 e
100A (<80 μm)	61.6 ± 3.9 a	14.6 ± 0.4 a	49.4 ± 4.7 a	128.3 ± 7.2 a
100B (80–180 μm)	186.3 ± 9.2 f	88.3 ± 9.6 d	177.0 ± 8.5 e	304.7 ± 9.7 d
100C (>180 μm)	354.2 ± 5.4 i	211.2 ± 1.4 f	337.4 ± 4.5 g	533.6 ± 12.2 h
50A/50B	124.1 ± 1.3 c	19.1 ± 0.4 ab	105.0 ± 0.6 c	263.2 ± 4.2 c
50A/50C	172.2 ± 4.8 e	18.9 ± 0.3 ab	105.6 ± 4.5 c	415.9 ± 8.6 e
50B/50C	261.1 ± 3.8 h	113.1 ± 3.0 e	242.4 ± 3.7 f	444.7 ± 4.9 f
25A/75B	151.9 ± 3.8 d	27.7 ± 1.3 bc	144.0 ± 5.4 d	287.3 ± 2.7 cd
75A/25B	101.1 ± 10.0 b	16.9 ± 0.4 a	74.2 ± 8.2 b	230.7 ± 23.3 b
25A/75C	252.8 ± 3.8 h	27.7 ± 0.1 c	250.8 ± 3.4 f	489.8 ± 9.0 g
75A/25C	113.1 ± 4.4 bc	16.5 ± 0.2 a	66.5 ± 1.6 b	299.5 ± 15.4 d

D[4,3]: average particle size which constitutes the bulk of the sample volume. D(10): maximum particle diameter below which 10% of the sample falls. D(50): maximum particle diameter below which 50% of the sample falls. D(90): maximum particle diameter below which 90% of the sample falls. Data are expressed as means ± standard deviation (SD) of duplicate assays. The values with the same letter in the same column do not present significant differences (at a significant level of $p < 0.05$).

According to Table 3, higher values of WHC are related to the compositions A and B, while lower values are found in mixtures with percentages of C higher than 50% (100C, 25A/75C and 50A/50C). In fact, there is a negative correlation between the WHC and D[4,3] $r = -0.74$ (significant at 99%). The relationship between WHC and the particle size was observed by other authors [10] and can be explained by the large surface area presented by finer flours.

Nevertheless, other studies were either not able to establish a correlation between particle size and WHC, or no decrease in WHC was found with reductions in particle size [14]. These differences could be related to both different compositions of particles with distinct sizes and to the morphology and particle size distribution, which was not evaluated in other studies. In this case, it is important to stress that the correlation of WHC with D(90) values was higher than with D[4,3], $r = -0.86$ (significant at 99.9%). This fact may indicate that the percentage of coarse fractions is what most affects this property. This could be explained by the fact that the finer fractions present similar particle sizes while the thicker fractions have greater differences between them. The swelling volume is highly correlated with WHC, D[4,3], and D(90) ($r = 0.80$, significant at 99%). Nevertheless, no correlation between particle size and WBC values was found, as observed by De la Hera et al. [15]. Thus, the larger values of WBC are related with fraction B followed by other individual fractions (pure A and pure C) and mixtures with the coarser fractions (B/C). On the other hand, the smaller values of WBC were observed in mixtures comprising A and B, except 50A/50B, which showed no significant difference compared to the control. Combinations of A and C presented intermediate values with no significant differences. Intermediate values of WBC were also shown by the control flour. In this case, the percentage of different fractions, their packing capacity and particle morphology seems to have a higher influence on this behaviour than the average particle size when using fine and coarse flours together in the centrifugal process.

3.2. Dough Rheology Properties

Dough rheology is a fundamental characteristic for cookie formulation so that if the dough is very soft or firm it is not easy to manipulate. The dough must be sufficiently cohesive to remain united during the different phases of processing and to be easily cut by the moulds [16]. In fact, an important finding from our rheology measurements was that values were obtained only for samples containing fraction A. The other doughs were extremely brittle and were impossible to laminate without breaking. From this result, it is possible to conclude that to make a cookie dough that is cohesive and laminable, it is necessary for it to contain a minimum proportion of reduced size particles. This conclusion was not found in other studies because those only considered pure fractions with a unique size, bigger or smaller, obtained by screening or by another grinding process.

In general, G' values were higher than G'' values (Table 3), showing that the elastic component is dominant over the viscous one, which suggests a solid elastic-like behaviour of all the doughs studied. This result was confirmed in values of tan delta that were all lower than 1.0, in agreement with other studies about cookies [8,17,18]. Between different doughs, smaller values of G' corresponded to those flours with a higher value of D[4,3], such as the control flour and the mixture 25A/75B. Meanwhile, the biggest values of G' were found in doughs with finer flours (with lower D[4,3]) with a higher percentage of fraction A. This corresponded with low values of D(10) (100A, 75A/25B, 75A/25C). This indicates that finer flours have a better packing quality, which results in more cohesive doughs. However, it is also important to note the larger values of G' obtained with mixtures of fractions A and C where C made up more than 50%. In those cases, it was very difficult to manipulate the doughs because of their weakness. These samples presented a different texture compared with other samples because they showed an anomalous rheological behaviour. These doughs were like those where it was not possible to measure the rheology parameters, but their breakability characteristics were less extreme.

Table 3. Maize flour, maize flour fractions and their combinations: hydration properties and cookie dough rheology.

	Hydration Properties			Dough Rheology		
	WHC	Swelling	WBC	G' ($\times 10^6$)	G'' ($\times 10^6$)	Tan Delta
Control flour	1.57 ± 0.04 [bcd]	2.09 ± 0.16 [abc]	1.28 ± 0.04 [de]	0.42 ± 0.02 [a]	0.25 ± 0.27 [a]	0.59 ± 0.54 [b]
100A (<80 µm)	1.85 ± 0.07 [fg]	2.29 ± 0.16 [d]	1.30 ± 0.01 [ef]	1.36 ± 0.27 [c]	0.26 ± 0.18 [a]	0.15 ± 0.04 [ab]
100B (80–180 µm)	1.87 ± 0.13 [g]	2.19 ± 0.01 [bcd]	1.47 ± 0.01 [g]	ND	ND	ND
100C (>180 µm)	1.48 ± 0.03 [bc]	2.20 ± 0.00 [cd]	1.30 ± 0.01 [ef]	ND	ND	ND
50A/50B	1.64 ± 0.10 [cde]	2.09 ± 0.14 [abc]	1.27 ± 0.04 [de]	0.72 ± 0.02 [ab]	0.12 ± 0.17 [a]	0.24 ± 0.35 [ab]
50A/50C	1.43 ± 0.09 [ab]	1.99 ± 0.01 [a]	1.19 ± 0.01 [bc]	1.20 ± 0.22 [c]	0.10 ± 0.13 [a]	0.10 ± 0.14 [ab]
50B/50C	1.83 ± 0.05 [fg]	2.27 ± 0.11 [cd]	1.33 ± 0.01 [f]	ND	ND	ND
25A/75B	1.74 ± 0.11 [efg]	2.18 ± 0.01 [abcd]	1.15 ± 0.02 [b]	0.27 ± 0.23 [a]	0.05 ± 0.30 [a]	0.25 ± 0.11 [ab]
75A/25B	1.69 ± 0.04 [def]	2.19 ± 0.02 [bcd]	1.05 ± 0.01 [a]	1.25 ± 0.18 [c]	0.06 ± 0.88 [a]	0.05 ± 0.08 [ab]
25A/75C	1.41 ± 0.02 [a]	2.00 ± 0.01 [ab]	1.20 ± 0.01 [c]	1.49 ± 0.17 [c]	0.08 ± 0.11 [a]	0.01 ± 0.00 [a]
75A/25C	1.56 ± 0.04 [abcd]	2.20 ± 0.01 [cd]	1.24 ± 0.04 [cd]	1.05 ± 0.15 [bc]	0.06 ± 0.90 [a]	0.14 ± 0.01 [ab]

ND: no development. WHC: water holding capacity. WBC: water binding capacity. Data are expressed as means ± SD of duplicate assays. The values with the same letter in the same column do not present significant differences (at a significance level of $p < 0.05$).

Regarding the viscous component (G"), it was not possible to find significant differences between the samples. Probably, this fact could be due to the excessive variability of the data, especially of the more fragile doughs. The same effect was observed in tan delta values, where some differences were found between control flour and the 25A/75C mixture, which were the most difficult samples to perform the rheology test on because of the cohesiveness. Even though previous studies reported correlations between rheological cookie values and hydration properties when flours or mixtures were used [18–20], in this case, no correlations were observed. This was due to the anomalous brittleness of samples with smaller proportions of fraction A and the importance between the particle size distribution and rheological behaviour.

3.3. Cookie Characteristics

Cookie characteristics are shown in Table 4. It is important to underline that it was not possible to make cookies with pure fractions B and C because the dough was very breakable (Figure 1). This brittle character was also observed in B/C mixtures (25B/75C, 50B/50C and 75B/25C) where, even though it was not possible to measure the rheological parameters, only the mixture 50B/50C was at least capable of being made. This fact could be explained because the rheological test required a thin layer of the dough, which was easily broken. These three flours (B, C, and B/C) were the ones that presented higher values of D(10), which means that it is necessary to add a small number of finer particles to the fractions. These particles take their place between the thick particles and increase the cohesiveness of the dough which allows the dough to be laminated. Furthermore, a mixture made using different particle sizes (B/C) seems to be more convenient than those made of only one size, which can be explained by the same effect of fine particles getting placed between coarse ones.

Table 4. Physical properties of cookies.

	Dimensions			Texture		Colour		
	Diameter (mm)	Spread Factor	Hardness (N)	Cookie Elastic Modulus (N/mm^2)	L^*	a^*	b^*	
Control flour	56.4 ± 0.40 c	7.48 ± 0.31 de	34.10 ± 1.22 cd	26.90 ± 1.52 ab	55.33 ± 1.97 c	5.59 ± 2.79 abc	18.33 ± 0.92 b	
100A (<80 μm)	43.1 ± 1.22 a	4.05 ± 0.24 a	50.84 ± 1.97 f	54.48 ± 2.73 d	78.45 ± 0.95 f	3.70 ± 1.40 a	21.41 ± 1.02 bcd	
100B (80–180 μm)	ND	ND	ND	ND	ND	ND	ND	
100C (>180 μm)	ND	ND	ND	ND	ND	ND	ND	
50A/50B	48.4 ± 0.22 bc	5.54 ± 0.34 bc	34.43 ± 1.94 cd	33.41 ± 3.13 bc	69.44 ± 1.27 e	5.47 ± 1.12 abc	21.37 ± 0.40 bcd	
50A/50C	55.1 ± 0.20 d	8.47 ± 0.20 ef	24.25 ± 0.37 b	27.15 ± 3.97 ab	61.20 ± 0.11 d	7.19 ± 0.76 bcd	19.14 ± 1.04 b	
50B/50C	54.9 ± 2.53 d	9.47 ± 1.02 f	25.98 ± 0.30 b	19.83 ± 0.45 a	51.11 ± 1.23 b	7.85 ± 1.40 cd	13.37 ± 1.04 a	
25A/75B	50.7 ± 0.07 c	6.31 ± 0.13 cd	36.24 ± 0.05 d	38.54 ± 1.38 c	69.84 ± 1.89 e	5.98 ± 0.55 abc	25.23 ± 1.12 d	
75A/25B	43.9 ± 1.22 a	4.53 ± 0.19 ab	42.82 ± 1.98 e	38.71 ± 10.61 c	75.66 ± 1.80 f	4.68 ± 0.60 ab	24.42 ± 1.32 cd	
25A/75C	63.0 ± 0.71 e	14.01 ± 1.31 g	17.93 ± 0.8 a	29.70 ± 5.37 abc	46.52 ± 3.10 a	9.71 ± 1.00 d	12.93 ± 2.13 a	
75A/25C	47.14 ± 0.79 b	5.10 ± 0.01 abc	33.24 ± 0.80 c	34.81 ± 1.74 bc	74.34 ± 2.63 f	5.19 ± 0.08 abc	21.06 ± 4.24 bc	

ND: no development. Data are expressed as means ± SD of duplicate assays. The values with the same letter in the same column do not present significant differences (at a significance level of $p < 0.05$).

Among the obtained cookies, no significant differences were found in final product moisture, and in all cases the values were less than 1%. It is also important to stress that the 25A/75C cookies showed the highest diameter and spread factor followed by those made with 50% of C (A/C and B/C). This fact seems to indicate that a higher percentage of coarse flour is favourable to releasing and spreading during baking, generating cookies of a greater diameter. This was also observed by Mancebo et al. [3] with rice flours, Rao et al. [7] with sorghum flour and Ai et al. [9] with dry bean powders. For their part, fractions with a higher percentage of A and, therefore, a lower average size (A, 75A/25B and 75A/25C) showed small spread factors; the next lowest spread factor was 50A/50B (with larger average size). In fact, 99.9% significant correlations were found between the diameter and D[4,3] ($r = 0.98$) and between the spread factor and D[4,3] ($r = 0.92$). Some authors affirm that flours with a lower hydration capacity produced cookies with higher diameters because they allowed excess water to dissolve the sugar, to reduce the initial viscosity of the doughs and to allow more expansion during baking [21,22]. Thus, in this study, significant correlations of 99% were found between WHC and the values of diameter ($r = -0.74$) and spread factor ($r = -0.72$), as found in similar studies [3,8]. In this way, it seems that the value of D[4,3] is a better indicator of dough expansion during baking and the final diameter, but D(10) shows a better possibility of obtaining cohesive doughs that can be laminated and cut without breaking.

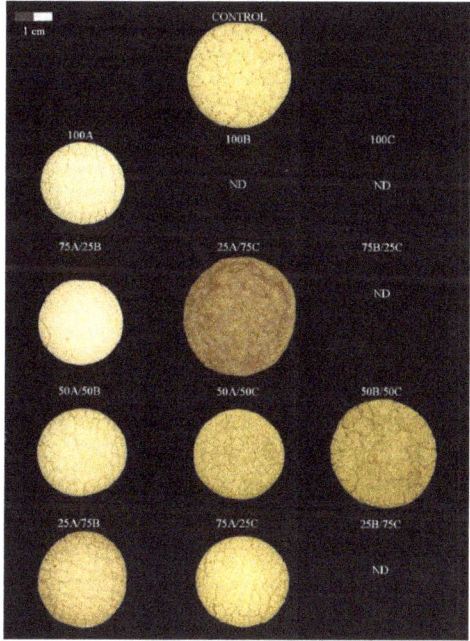

Figure 1. Image from cookies made with different particles sizes of white maize flour.

Regarding the texture of the cookies, a significant correlation coefficient of $r = 0.88$ was found in all cases but at differing significance levels: 99.9% significance between hardness, diameter and spread factor; 99% between the elastic modulus and diameter and 95% between the elastic modulus and the spread factor). Therefore, cookie dimensions showed a stronger correlation with hardness than with the gradient obtained by the curve. Thus, harder cookies presented both a lower diameter and spread factor (100A and 75A/25B). On the other hand, less hard cookies showed higher spread factors (25A/75C) followed by those mixtures with A/C and B/C which presented values greater than all the other samples. Mancebo et al. [3] described how cookies with higher spread factors showed a lower peak force in textural analysis, but we found no correlation whatsoever between these two parameters for any of the cookies we analysed. In addition to the dimensions of the cookies, these differences may be related to their internal structure, the particle size of the flours and their compaction capacity. Parameters of texture and indicators of flour particle size showed a better correlation than with cookie dimensions, which demonstrates that as the particle size decreases, the hardness of the cookies increases, in agreement with the observations of similar studies [7,9]. In the case of D[4,3], the correlation with hardness also presented a significant correlation of $r = -0.87$ (99% significance). One observed correlation, between hardness and D(90), stands out ($r = -0.93$, significant at 99.9%), as it is higher even than the ones observed between hardness and D(50) ($r = -0.72$, significant at 99%) and between hardness and D(10) ($r = -0.61$, significant at 95%). Therefore, it seems that what most reduced the hardness of the cookies was an increase in the percentage of thicker particles, even though it is necessary to remember the importance of having a certain percentage of smaller particles to allow the formation of the cookies.

Regarding colour, important differences were found among the values of L^*. Cookies with a larger percentage of A (100% or 75%) were lighter in colour and presented lower spread factors. Meanwhile, cookies made with a larger percentage of C (25A/75C) showed darker colour (small values of L^*) followed by those made with 50% of C. With regard to a^* and b^* values, only those samples with extreme mixes stand out: sample A, which had a lower particle size, diameter and spread factor, presented the smallest values of a^*, while the sample 25A/75C, which had a higher particle size, spread factor and diameter, showed the highest values of a^* and smallest of b^*. Mancebo et al. [3] also observed the smallest values of L^* and b^* for cookies with higher spread. This could be because of a higher diameter and a lower thickness of cookies, which produces high temperatures in a short time within the dough. Thus, temperature increases led to more caramelization and Maillard reactions, which are the main causes of final cookie colour [23]. In fact, significant correlations of 99.9% were found between the spread factor of the cookies and the parameters L^* ($r = -0.91$), a^* ($r = -0.87$) and b^* ($r = -0.81$).

4. Conclusions

The final characteristics of cookies can be influenced by the properties of the ingredients used in their elaboration such as the particle size of the flour, the sugar content and the fat used in formulation. There are no references about how both of these last factors can affect the quality of the cookies, which may be an opportunity for future studies. However, previous studies proved that a higher average particle size favours spread factor and decreases cookie hardness. Nevertheless, this study proves for the first time that it is necessary to also have a certain percentage of fine particles, which, being placed between coarser particles in the dough, give rise to a higher dough cohesiveness. Otherwise, doughs are excessively fragile in the laminating process and cookies are impossible to be made. These results must be considered in future studies about gluten-free cookies and in food development in the industry.

Author Contributions: Conceptualization, M.G.; methodology, M.G and M.S.; validation, M.G and M.S; formal analysis, M.B.; investigation, M.B. and M.S.; data curation, M.G and M.B.; writing—original draft preparation, M.B.; writing—review and editing, M.B. and M.G.; visualization, M.G and M.S.; supervision, M.G and M.S.; project administration, M.G.; funding acquisition, M.G.

Funding: The authors acknowledge the financial support of the Spanish Ministry of Economy and Competitiveness (Project AGL2014-52928-C2) and the European Regional Development Fund (FEDER).

Acknowledgments: The authors are also grateful to Molendum Ingredients for supplying the maize flour.

Conflicts of Interest: The authors declare no conflict of interest.

References

1. Crowe, S.E. Celiac disease. In *Nutritional and Gastrointestinal Disease*; Human Press Inc.: Totowa, NJ, USA, 2008; pp. 123–147.
2. Pareyt, B.; Delcour, J.A. The role of wheat flour constituents, sugar, and fat in low moisture cereal-based products: A review on sugar-snap cookies. *Crit. Rev. Food Sci. Nutr.* **2008**, *48*, 824–839. [CrossRef] [PubMed]
3. Mancebo, C.M.; Picon, J.; Gómez, M. Effect of flour properties on the quality characteristics of gluten free sugar-snap cookies. *LWT* **2015**, *64*, 264–269. [CrossRef]
4. Torbica, A.; Hadnadev, M.; Hadnadev, T.D. Rice and buckwheat flour characterization and its relation to cookie quality. *Food Res. Int.* **2012**, *48*, 277–283. [CrossRef]
5. Rai, S.; Kaur, A.; Singh, B. Quality characteristics of gluten free cookies prepared from different flour combinations. *J. Food Sci. Technol.* **2014**, *51*, 785–789. [CrossRef] [PubMed]
6. Barak, S.; Mudgil, D.; Khatkar, B.S. Effect of flour particle size and damaged starch on the quality of cookies. *J. Food Sci. Technol.* **2014**, *5*, 1342–1348. [CrossRef] [PubMed]

7. Rao, B.D.; Anis, M.; Kalpana, K.; Sunooj, K.V.; Patil, J.V.; Ganesh, T. Influence of milling methods and particle size on hydration properties of sorghum flour and quality of sorghum biscuits. *LWT* **2016**, *67*, 8–13.
8. Mancebo, C.M.; Rodriguez, P.; Gómez, M. Assessing rice flour-starch-protein mixtures to produce gluten free sugar-snap cookies. *LWT* **2016**, *67*, 127–132. [CrossRef]
9. Ai, Y.F.; Jin, Y.N.; Kelly, J.D.; Ng, P.K.W. Composition, functional properties, starch digestibility, and cookie-baking performance of dry bean powders from 25 Michigan-grown varieties. *Cereal Chem.* **2017**, *94*, 400–408. [CrossRef]
10. Protonotariou, S.; Batzaki, C.; Yanniotis, S.; Mandala, I. Effect of jet milled whole wheat flour in biscuits properties. *LWT* **2016**, *74*, 106–113. [CrossRef]
11. Barrera, G.N.; Pérez, G.T.; Ribotta, P.D.; León, A.E. Influence of damaged starch on cookie and bread-making quality. *Eur. Food Res. Technol.* **2007**, *225*, 1–7. [CrossRef]
12. AACC International. *AACC International Approved Methods, AACCI Method 56-30.01, Water Hydration Capacity of Protein Materials*; AACC International: St. Paul, MN, USA, 1999.
13. AACC International. *AACC International Approved Methods, AACCI Method 44-15.02, Moisture—Air-Oven Methods*; AACC International: St. Paul, MN, USA, 1999.
14. Ahmed, J.; Al-Attar, H.; Arfat, Y.A. Effect of particle size on compositional, functional, pasting and rheological properties of commercial water chestnut flour. *Food Hydrocoll.* **2016**, *52*, 888–895. [CrossRef]
15. De la Hera, E.; Talegón, M.; Caballero, P.; Gómez, M. Influence of maize flour particle size on gluten-free breadmaking. *J. Sci. Food Agric.* **2013**, *93*, 924–932. [CrossRef] [PubMed]
16. Gujral, H.S.; Mehta, S.; Samra, I.S.; Goyal, P. Effect of wheat bran, coarse wheat flour and rice flour on the instrumental texture of cookies. *Int. J. Food Prop.* **2003**, *6*, 329–340. [CrossRef]
17. Lee, S.; Inglett, G.E. Rheological and physical evaluation of jet-cooked oat bran in low calorie cookies. *Int. J. Food Sci. Technol.* **2006**, *41*, 553–559. [CrossRef]
18. Mancebo, C.M.; Rodriguez, P.; Martínez, M.M.; Gómez, M. Effect of the addition of soluble (nutriose, inulin and polydextrose) and insoluble (bamboo, potato and pea) fibres on the quality of sugar-snap cookies. *Int. J. Food Sci. Technol.* **2018**, *53*, 129–136. [CrossRef]
19. Zhang, Q.; Zhang, Y.; Zhang, Y.; He, Z.H.; Peña, R.J. Effects of solvent retention capacities, pentosan content, and dough rheological properties on sugar snap cookie quality in Chinese soft wheat genotypes. *Crop. Sci.* **2007**, *47*, 656–664. [CrossRef]
20. Inglett, G.E.; Chen, D.; Liu, S.X. Physical properties of gluten-free sugar cookies made from amaranth-oat composites. *LWT* **2015**, *63*, 214–220. [CrossRef]
21. Yamazaki, W.T. The concentration of a factor in soft wheat flours affecting cookie quality. *Cereal Chem.* **1955**, *32*, 26–37.
22. Hoseney, R.C.; Rogers, D.E. Mechanism of sugar functionality in cookies. In *The Science of Cookie and Cracker Production*; Faridi, H., Ed.; Avi: New York, NY, USA, 1994; pp. 203–226.
23. Chevallier, S.; Colonna, P.A.; Della Valle, G.; Lourdin, D. Contribution of major ingredients during baking of biscuit dough systems. *J. Cereal Sci.* **2000**, *3*, 241–252. [CrossRef]

© 2019 by the authors. Licensee MDPI, Basel, Switzerland. This article is an open access article distributed under the terms and conditions of the Creative Commons Attribution (CC BY) license (http://creativecommons.org/licenses/by/4.0/).

Article

Tempering Improves Flour Properties of Refined Intermediate Wheatgrass (*Thinopyrum intermedium*)

Catrin Tyl, Radhika Bharathi, Tonya Schoenfuss and George Amponsah Annor *

Department of Food Science and Nutrition, University of Minnesota, 1334 Eckles Avenue, Saint Paul, MN 55108, USA
* Correspondence: gannor@umn.edu; Tel.: +1-612-624-3201

Received: 1 July 2019; Accepted: 7 August 2019; Published: 10 August 2019

Abstract: Progress in breeding of intermediate wheatgrass (*Thinopyrum intermedium*), a perennial grain with environmental benefits, has enabled bran removal. Thus, determination of optimum milling conditions for production of refined flours is warranted. This study explored the effect of tempering conditions on intermediate wheatgrass flour properties, namely composition, color, solvent retention capacity, starch damage, and polyphenol oxidase activity. Changes in flour attributes were evaluated via a 3 × 3 × 2 factorial design, with factors targeting moisture (comparing un-tempered controls to samples of 12% and 14% target moisture), time (4, 8, and 24 h), and temperature (30 and 45 °C). All investigated parameters were significantly affected by target moisture; however, samples tempered to 12% moisture showed few differences to those tempered to 14%. Similarly, neither tempering time nor temperature exerted pronounced effects on most flour properties, indicating water uptake was fast and not dependent on temperature within the investigated range. Lactic acid retention capacity significantly correlated with ash ($r = -0.739$, $p < 0.01$), insoluble dietary fiber ($r = -0.746$, $p < 0.01$), polyphenol oxidase activity ($r = -0.710$, $p < 0.01$), starch content ($r = 0.841$, $p < 0.01$), and starch damage ($r = 0.842$, $p < 0.01$), but not with protein ($r = 0.357$, $p > 0.05$). In general, tempering resulted in flour with less bran contamination but only minor losses in protein.

Keywords: flour refinement; intermediate wheatgrass; sustainability; tempering; bran

1. Introduction

Consumers are increasingly interested in sustainably produced food ingredients, and alternative agricultural models have emerged to emphasize factors such as diversity and environmental concerns [1]. Perennial grains have the potential to become serious contenders in the cereal market place, provided that their quality matches the requirements of manufacturers as well as expectations of consumers [2]. The motivation for their use stems from their efficient use of water, fertilizers, and soil nutrients because of their extended root systems [3]. For instance, it has been shown that cultivation of the perennial grain intermediate wheatgrass (*Thinopyrum intermedium*, IWG) dramatically lowers environmental strains such as nitrate leaching [4].

Aside from sustainability aspects, the nutritional profile of IWG is one of its advantages for food use, due to protein and dietary fiber contents that surpass those of many other cereal grains, including wheat [5–7]. However, the protein makeup of IWG differs from wheat, specifically by being deficient in high-molecular weight glutenins [6,7], with deleterious effects on dough elasticity and gas holding properties [6,8]. The short domestication history of this perennial grain is reflected in a rather narrow endosperm, which, however, has been successfully increased by breeding efforts. Even though IWG kernel width is still significantly lower than for wheat [8], refinement of newer breeding materials is possible, which has been shown to retain most of the protein, while reducing especially the insoluble fiber portion [8]. In order to explore additional uses for IWG as a standalone food ingredient, processing conditions to better separate the endosperm from the bran fraction need to be investigated. Many cereal applications involve the use of flour, and, thus, milling is one of the most important processes to improve end-use characteristics of cereals. One strategy for doing so is by altering the moisture contents of kernels prior to milling. This tempering operation can increase milling efficiency [9] and affect parameters relevant for final application of the resulting flour. As there are currently no tempering conditions established for IWG, this study explored the effects of kernel moisture, tempering temperature, and tempering time on chemical characteristics of IWG. Specifically, IWG kernels were subjected to different tempering conditions in a $3 \times 3 \times 2$ factorial design. Kernels were incubated at either 30 or 45 °C for 4, 8, or 24 h. Nontempered kernels were compared to those where water had been added to achieve either 12% or 14% target moisture. Flour samples were analyzed for composition (ash, protein, starch, and dietary fiber), polyphenol oxidase (PPO) activity, color, solvent retention capacity (SRC), and starch damage. Initial work on IWG incorporation into baked goods utilized it in whole-grain form [6]. In a previous study, properties of doughs made of completely, partially, and unrefined IWG were compared, as well as the influence of addition of commercial dough conditioners on these properties in relation to refinement [8]. It was noticed that dough conditioners had more pronounced effects on dough properties in completely refined IWG [8]. In addition, crumb characteristics of breads were improved when ascorbic acid was part of the recipes, but only when flours had been completely refined [10]. However, the flours used in these experiments had not been tempered before milling; therefore, additional refinement due to better bran separation may have yielded further enhanced properties. Thus, this study is part of a wider reaching effort to improve IWG functionality for baking applications to facilitate its incorporation into a wide array of products with desirable properties. The overall objective of this wider study is to optimize the tempering conditions of IWG for end use applications. For this study, we focused on the effects of tempering on the chemical composition, color, solvent retention capacity, starch damage, and polyphenol oxidase activity.

2. Materials and Methods

IWG variety MN1503 was grown and harvested in Becker, Minnesota, USA. All chemicals used were of reagent grade or higher. Tempering as well as chemical analyses were carried out according to American Association of Cereal Scientists International (AACCI) official methods [11] unless otherwise noted.

Samples were tempered in duplicates at 4, 8, or 24 h at either 30 or 45 °C. Kernels were either used at intrinsic moisture (i.e., no water added) as controls or tempered to 12% ± 0.5% or 14% ± 0.5% target moisture. Tempering was performed by weighing 50 g of IWG kernels into air-sealed glass jars, to which the required amount of tempering water was added, hand mixed using a spatula for 60 s, shaken for another 60 s to ensure equal distribution of water, and tempered in an incubator at each condition with fixed relative humidity of 50%. The amount of tempering water to be added to each tempering condition was calculated as per AACCI method 26-10.02.

At the end of each tempering condition, duplicates of 1 g of kernels were removed for whole kernel moisture analysis via AACCI method 44-16.01, and remaining kernels were milled using a Brabender Quadrumat Junior mill (Model12-02-000, C.W. Brabender Instruments, Hackensack, NJ, USA). To avoid flour losses, bran was sifted manually using a 250 µm sieve plate. Flour with particle sizes >250 µm was collected as bran and flour with particle sizes <250 µm as endosperm. Moisture contents of the freshly milled flour were determined on an Ohaus MB45 (Parsippany, NJ, USA).

Flours were analyzed for ash (AACCI method 08-01.01), protein (AACCI method 46-30.01) on a Leco FP 828 (Leco 165 Corporation, St. Joseph, MI, USA), color (AACCI method 14.22.01) Chroma Meter CR-221 (Minolta Camera Co., Osaka, Japan), dietary fiber (AACCI method 32-07.01), SRC (AACCI method 56-11.02), total starch (AACCI method 76-13.01), and starch damage (AACCI method 76-31.01). For dietary fiber, total and damaged starch, test kits from Megazyme (Wicklow, Ireland) were used. The activity of polyphenol oxidases (PPOs) were determined in flour based on a previously reported procedure [12]. Briefly, 200 mg of flour were used to assess oxidation of 3,4-dihydroxy-L-phenylalanine by measuring A_{475} of its conversion product on a UV1800 (Shimadzu, Columbia, MD). The substrate was dissolved in 50 mM 3-morpholinopropane-1-sulfonic acid buffer adjusted to pH 6.5 with HCl, containing 0.02% Tween-20. Samples were shaken for 1 h on a Fisher Pulsing Vortex mixer operated at a speed setting of 2500 (Fisher Scientific, Waltham, MA, USA). Enzyme activity is reported as absorbance change ΔA_{475} min^{-1} g^{-1} flour dry basis, as outlined previously [12].

Three-way analysis of variance (ANOVA) was performed in R (version 3.1.0, R Core Team, 2013) using target moisture, tempering time, and temperature as factors. All analyses were carried out at least in duplicate and are reported as means on a dry basis ± standard deviation. Differences among means were considered significant if $p < 0.05$ and were compared within each factor. Only significant differences and correlations are reported. One-way ANOVAs followed by least significance difference (LSD) tests were run in R to evaluate differences according to target moisture and tempering time. Paired t-tests were conducted in Microsoft Excel® (2013) to evaluate differences according to tempering temperature. Pearson-type correlation coefficients and principal component analyses were done in R to summarize relationships among the data.

3. Results and Discussion

3.1. Flour and Kernel Moisture Uptake During Tempering

Target moisture exerted the greatest effect on kernel as well as flour moisture, in comparison to temperature and time (Table 1). For all tempering conditions, nontempered samples had significantly lower flour and kernel moisture than samples tempered to 12% or 14% moisture. The fact that these differences were already significant after 4 h indicated rapid water uptake. The observed differences in the moisture contents of the nontempered samples shown in Table 1 were due to the fact that kernel moisture was determined prior to each tempering experiment, even though kernels were from the same batch. This was necessary because not all tempering experiments could be performed on the same day. Incubation time only affected kernel moisture values obtained for nontempered samples at 30 °C and IWG tempered to 12% target moisture at 45 °C. The length of tempering did not cause significant differences in flour moisture of samples tempered to either 12% or 14% target moisture. In wheat, tempering times are often optimized for cultivars, and lengths of 1–24 h are common [13]. However, water uptake is not uniform across grain tissue layers because of the differences in water permeability [13]. Ultimately, the goal of tempering is to facilitate separation of endosperm and bran by plasticizing the former while concomitantly toughening the latter.

For samples with 14% target moisture, more variability between replicate tempering treatments was observed, and not all samples reached the intended 14% level, which resulted in the differences between 12% and 14% target moisture as not being significant. Regardless of tempering conditions, flour moisture was lower than the corresponding kernel moisture. This could be due to the loss of moisture from the flour during the milling process. As seen in Table 1, target moisture showed significant differences in the flour moisture values recorded for all samples, excluding tempering conditions 30 °C 24 h 12% and 30 °C 24 h 14% wherein no significant differences were observed. Tempering temperature did not significantly affect kernel moisture except for samples tempered for 8 h to 12% or 14% target moisture. Moreover, temperature did not significantly affect flour moisture of any samples tempered to 12% and 14% target moisture.

Table 1. Intermediate wheatgrass kernel and flour moisture as affected by tempering conditions.

30 °C	Incubation Time					
	4 h		8 h		24 h	
Target Moisture (%)	Kernel	Flour	Kernel	Flour	Kernel	Flour
NT [1]	8.85 ± 0.05 c	8.94 ± 0.04 C*	9.00 ± 0.05 c	9.22 ± 0.15 C	9.22 ± 0.02 c	8.47 ± 0.02 B
12	11.88 ± 0.27 b	10.68 ± 0.07 B	12.10 ± 0.06 b*	10.09 ± 0.07 B	11.51 ± 0.49 b	10.7 ± 0.45 A
14	13.93 ± 0.35 a	12.06 ± 0.20 A	14.07 ± 0.01 a*	12.02 ± 0.23 A	13.45 ± 0.48 a	11.44 ± 0.20 A
45 °C	4 h		8 h		24 h	
Target Moisture (%)	Kernel	Flour	Kernel	Flour	Kernel	Flour
NT	8.72 ± 0.03 b	8.59 ± 0.02 C*	9.02 ± 0.25 c	9.09 ± 0.14 C	9.09 ± 0.03 c	8.26 ± 0.10 C
12	12.62 ± 0.06 a	11.21 ± 0.11 B	12.01 ± 0.06 b*	11.05 ± 0.06 B	11.76 ± 0.05 b	11.17 ± 0.03 B
14	13.42 ± 0.54 a	12.32 ± 0.09 A	13.61 ± 0.01 a*	12.06 ± 0.14 A	13.45 ± 0.01 a	12.24 ± 0.03 A

[1] NT, nontempered control flours. Different letters represent significant ($p < 0.05$) differences among samples incubated at the same temperature for the same length of time, but to different target moistures. Lowercase letters denote differences for kernels; uppercase letters indicate differences among flour samples. Asterisks represent differences due to incubation temperature.

3.2. Color

Similar to flour and kernel moisture, the target moisture was the most impactful factor on color (Table 2). All samples tempered to 12% or 14% target moisture had significantly lower *a* values, and most also had significantly higher *L* and *b* values. Thus, tempering resulted in flours that were lighter, less reddish/brown, but more yellow than nontempered controls. These results indicate successive changes in flour refinement with increased target moisture, as darker, more reddish/browner but less yellow samples contain more bran. No significant differences according to tempering length or temperature were detected.

Table 2. Color of intermediate wheatgrass flour subjected to different tempering conditions.

L	Incubation Time/Temperature					
	4 h		8 h		24 h	
Target Moisture (%)	30 °C	45 °C	30 °C	45 °C	30 °C	45 °C
NT	85.90 ± 0.63 b	85.35 ± 0.86 b	85.75 ± 0.43 b	86.25 ± 0.31	84.70 ± 0.28 b	86.11 ± 0.59 b
12	88.34 ± 0.11 a	87.89 ± 0.22 a	88.09 ± 0.17 a	88.03 ± 0.26	86.84 ± 1.25 ab	87.76 ± 0.35 a
14	88.57 ± 0.25 a	88.25 ± 0.14 a	88.72 ± 0.01 a	88.51 ± 0.22	88.98 ± 0.04 a	88.14 ± 0.07 a
a						
NT	−6.19 ± 0.06 a	−5.99 ± 0.00 a	−5.86 ± 0.04 a	−6.07 ± 0.16 a	−5.52 ± 0.35 a	−6.10 ± 0.11 a
12	−6.93 ± 0.05 b	−7.05 ± 0.07 b	−6.83 ± 0.07 b	−6.93 ± 0.08 b	−6.58 ± 0.19 b	−6.78 ± 0.07 b
14	−7.15 ± 0.11 b	−7.14 ± 0.02 b	−7.06 ± 0.11 b	−7.23 ± 0.06 b	−7.08 ± 0.13 b	−7.09 ± 0.05 c
b						
NT	19.24 ± 0.69 b	18.95 ± 0.11 c	18.90 ± 0.29 c	19.33 ± 0.23 c	18.27 ± 0.65	19.24 ± 0.56 b
12	20.32 ± 0.06 ab	20.56 ± 0.03 b	20.35 ± 0.31 b	20.69 ± 0.48 b	19.58 ± 1.14	20.51 ± 0.34 ab
14	21.32 ± 0.33 a	21.34 ± 0.36 a	21.40 ± 0.07 a	21.84 ±0.31 a	21.32 ± 0.13	21.29 ± 0.35 a

[1] NT, nontempered control flours. Different lowercase letters represent significant ($p < 0.05$) differences among samples incubated at the same temperature and for the same length, but tempered to different target moistures (i.e., across columns). Target moisture did not significantly affect *L* values of samples incubated at 45 °C for 8 h and *b* values of samples incubated at 30 °C for 24 h.

3.3. Flour Composition

3.3.1. Ash, Insoluble Dietary Fiber, and Protein

Cereal bran in general, and IWG bran in particular, is richer in ash and dietary fiber (particularly insoluble dietary fiber) than refined flour [8]. Thus, higher ash values are often taken as a proxy for bran contamination of endosperm [9]. Target moisture significantly affected ash and insoluble dietary fiber in a similar way as observed for color (Table 2): all nontempered control flours contained significantly more ash and insoluble fiber than samples tempered to either 12% or 14% target moisture. In contrast, refinement did not always lead to protein loss when considering values on a dry basis. It has been shown before that IWG surpasses many other cereals, including wheat, in protein content [5–8], and this characteristic is maintained over the refinement process. From a nutritional perspective, a target moisture of 12% may be more desirable than 14% since it often results in more insoluble fiber, while not differing in ash contents to those samples. However, future studies would need to evaluate if dough rheological properties are different between such samples. It was previously observed that dough viscoelasticity of IWG is detrimentally affected by the presence of bran [8]. The effects of tempering time and temperature on the compositional attributes reported in Table 3 were minor compared to target moisture. However, tempering for 8 h at 45 °C resulted in lower ash contents than tempering for 4 or 24 h.

Ash and insoluble dietary fiber contents displayed significant, negative correlations with L (−0.849 and −0.909, respectively; $p < 0.01$ for both) and b values ($r = -0.824$ and −0.887, respectively, $p < 0.01$ for both) and significant and positive correlations with a values ($r = 0.898$ and 0.934, respectively, $p < 0.01$ for both). The negative relationship between lightness and ash contents has been reported numerous times for other cereal flours, notably wheat [14] and rye milling fractions [15]. However, for rye it was observed that lighter flours were not only less red, but also less yellow, because of the lower brownness [15]. This discrepancy to IWG may be related to differences in carotenoid contents since these would contribute to yellowness more than to redness. The contents of zeaxanthin, and especially lutein, in IWG exceeded ranges reported for many other cereals such as wheat [7] and are the reason for high b values (Table 2) in comparison to flours from, for example, wheat [16]. In contrast to the relation between color values and ash, correlations between L, a, and b to protein contents were not significant. Protein did, however, exhibit a positive correlation with ash ($r = 0.843$, $p < 0.01$) but not with insoluble dietary fiber.

Table 3. Ash, insoluble dietary fiber, and protein contents of intermediate wheatgrass flour (g/100 g, dry basis) as affected by tempering conditions.

		\multicolumn{6}{c}{Incubation Time/Temperature}				
	Target Moisture (%)	4 h		8 h		24 h	
		30 °C	45 °C	30 °C	45 °C	30 °C	45 °C
Ash	NT [1]	0.82 ± 0.00 aB	0.80 ± 0.04 a	0.90 ± 0.01 aB	0.79 ± 0.01 a	1.02 ± 0.05 aA	0.77 ± 0.03 a
	12	0.69 ± 0.01 b	0.66 ± 0.01 bA	0.70 ± 0.04 b	0.59 ± 0.01 bB	0.71 ± 0.02 b*	0.63 ± 0.02 bA*
	14	0.68 ± 0.04 b	0.67 ± 0.02 b	0.71 ± 0.03 b*	0.65 ± 0.03 b*	0.70 ± 0.03 b	0.64 ± 0.01 b
Insoluble dietary fiber	NT	5.81 ± 0.39 a	7.45 ± 0.12 aA	6.83 ± 0.55 a	6.93 ± 0.24 aA	8.50 ± 1.18 a	6.46 ± 0.31 aB
	12	4.39 ± 0.51 b	3.90 ± 0.22 b	4.07 ± 0.50 b	4.31 ± 0.41 b	4.55 ± 0.43 b	3.65 ± 0.17 b
	14	2.37 ± 0.02 cB	3.97 ± 0.28 bA	2.79 ± 0.96 bB*	4.00 ± 0.00 bA*	4.69 ± 0.12 bA	3.22 ± 0.18 bB
Protein	NT	19.25 ± 0.10	19.45 ± 0.06 aAB	19.35 ± 0.14	19.57 ± 0.08a A	19.01 ± 0.06 b	19.28 ± 0.03 B
	12	19.34 ± 0.19	19.19 ± 0.02 b	19.39 ± 0.05	19.21 ± 0.03 b	19.37 ± 0.04 a	19.26 ± 0.02
	14	19.04 ± 0.05	19.11 ± 0.10 b	19.48 ± 0.29	18.99 ± 0.06 c	19.16 ± 0.06 b	19.27 ± 0.09

[1] NT, nontempered control flours. Different lowercase letters signify differences among samples incubated according to target moisture (i.e., across columns). Different uppercase letters denote differences among samples according to tempering time, and asterisks represent differences due to incubation temperature.

3.3.2. Starch and Damaged Starch

The ash and insoluble fiber reductions achieved via tempering were reflected in significantly higher starch contents with higher target moisture (Figure 1a,b). This was the case at both tempering temperatures and for all incubation times except for 4 h at 30 °C, where the difference between nontempered control flour and 12% target moisture was not significant. As observed for color and the other evaluated constituents (Table 3), few differences between tempering length and temperature were significant. The starch values observed here were higher than in previous studies [5–7], which, however, had been measured on whole-grain IWG from different breeding stages.

The increases in starch contents coincided with increases in starch damage, which ranged from 3.04% ± 0.10% to 8.78% ± 0.24% (Figure 1a,b). While the values at the low range were in line with previous reports [17], and comparable to some soft wheat varieties [18,19] or rye [15], the values at the upper range were closer to medium-hard [20] or even hard wheat varieties [21]. Different from the compositional attributes shown in Table 3, tempering temperature caused more differences among samples. Except for samples incubated for 24 h at 12% target moisture, all tempered samples had significantly higher starch damage when incubated at 45 °C. Starch and starch damage exhibited significant, positive correlations with L (r = 0.760 and 0.692, respectively; p < 0.01 for both) and b (r = 0.807 and 0.809, respectively; p < 0.01 for both) and significant, negative correlations with a (r = −0.817 and −0.767, respectively, p < 0.01 for both). These relationships essentially indicate that all these parameters assess the extent of IWG refinement. While tempering may reduce the extent of starch damage by softening the kernels, some studies have found more starch damage in refined than whole wheat flours and attributed this to starch dilution by bran in whole wheat [19]. At high extents, starch damage can negatively influence product properties (e.g., lower functionality and sensory properties [22]), but below a certain threshold it may also be beneficial for product quality [23].

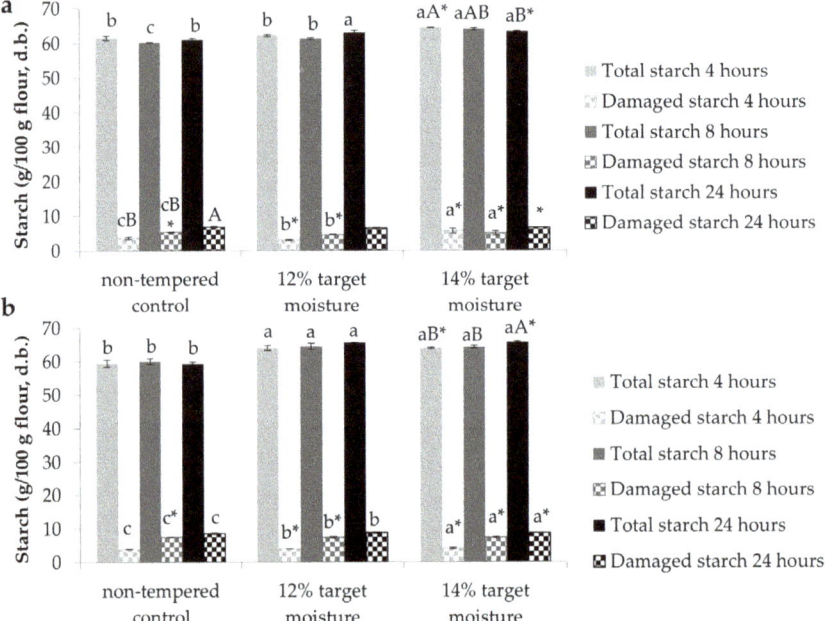

Figure 1. Starch and damaged starch contents in (**a**) samples incubated at 30 °C and (**b**) samples incubated at 45 °C. Lowercase letters reflect differences according to target moisture, uppercase letters illustrate differences due to incubation time, and asterisks represent differences due to incubation temperature.

3.4. Polyphenol Oxidase (PPO) Activity

Aside from ash contents, PPO activity is often cited as another indicator of bran contamination [9]. PPO activity can lead to discoloration and specks in products such as noodles. The values observed in our study were similar to those reported for PPO activity in wheat flour [12] and follow the same pattern as color, ash, and insoluble fiber data: nontempered controls had significantly higher PPO activity than samples of 12% or 14% target moisture, while incubation time and temperature had comparably minor effects (Figure 2a,b). Thus, similar to previous studies [9], PPO activity reflected refinement increases brought about by tempering and, consequently, was highly, positively correlated with a values ($r = 0.895$, $p < 0.01$), ash ($r = 0.843$, $p < 0.01$), and insoluble dietary fiber ($r = 0.921$, $p < 0.01$) and negatively correlated with starch ($r = -0.727$, $p < 0.01$), L ($r = -0.890$, $p < 0.01$), and b values ($r = -0.808$, $p < 0.01$). However, there was no significant correlation between PPO activity and protein contents (Table S1) because bran reduction did not decrease protein contents to a similar extent than it reduced PPO activity (Table 3). The low PPO activity in combination with high b values of the tempered IWG flours makes them attractive candidates for product applications such as noodles.

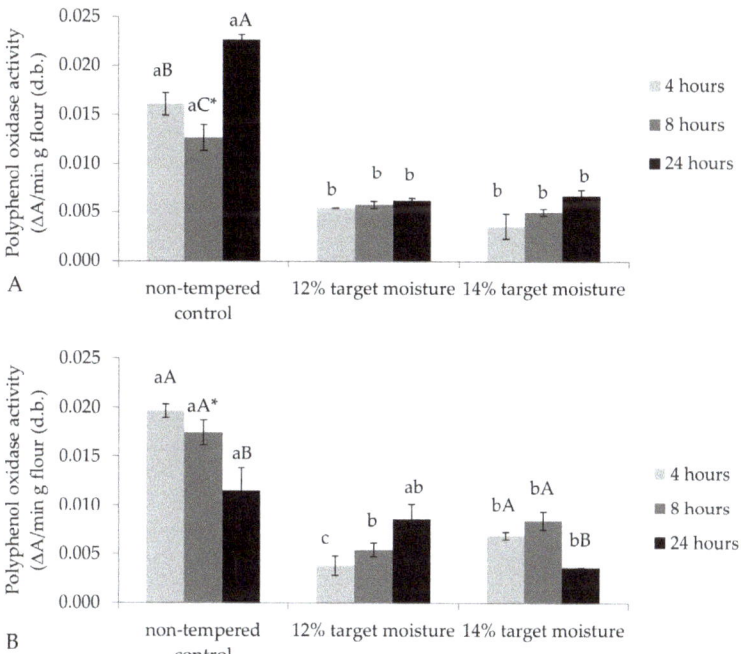

Figure 2. Polyphenol oxidase activities in (**A**) samples incubated at 30 °C and (**B**) samples incubated at 45 °C. Lowercase letters reflect differences according to target moisture, uppercase letters illustrate differences due to incubation time, and asterisks represent differences due to incubation temperature.

3.5. Solvent Retention Capacity

Flour swelling in lactic acid is related to gluten formation and is, thus, related to protein quality [24]. For samples tempered at 30 °C for 4 or 24 h, as well as samples tempered at 45 °C for 4 or 8 h, nontempered controls had significantly lower lactic acid SRCs than samples of 12% or 14% target moisture (Table 4). While the lactic acid SRC values were lower than for hard wheat, they were higher than previously reported for refined IWG [17]. This indicates that when the extent of refinement is increased, IWG flours develop stronger protein networks, which would be beneficial for the manufacture of products that require gas holding properties. The fact that several tempering conditions resulted in flour with significantly increased lactic acid SRC may be related to lower bran contents in these flours and, thus, less interference with protein network formation. In a previous study on IWG, bran addition to refined IWG resulted in changes in protein secondary structure distributions, with refined IWG having significantly fewer β-sheets, but more β-turns, than whole IWG [8]. Such changes reflect water redistribution, due to bran constituents, and a less hydrated protein network in whole IWG according to the loop and train model [25]. Thus, a higher lactic acid SRC may be indicative of better viscoelastic properties, which need to be evaluated in future studies. Lactic acid SRC was positively correlated with L ($r = 0.805$, $p < 0.01$), b ($r = 0.831$, $p < 0.01$), starch content ($r = 0.841$, $p < 0.01$), and also starch damage ($r = 0.842$, $p < 0.01$); accordingly, negative correlations were present with a ($r = -0.837$, $p < 0.01$), ash ($r = -0.739$, $p < 0.01$), insoluble dietary fiber ($r = -0.746$, $p < 0.01$), and PPO activity ($r = -0.710$, $p < 0.01$). Several previous studies have shown correlations between lactic acid SRC and protein or gluten content [21]; however, correlations with protein content were not always present when different wheat varieties were compared to each other [26]. Since, overall, gluten network formation, and consequently baking quality, is affected by the interplay among flour polymers, most importantly gluten-forming proteins, damaged starch, and arabinoxylans [24]. There are also reports that indicates that lactic acid SRC correlations with starch content may not always be significant and can be positive [27,28]. The positive correlations of lactic acid SRC with L, b, and starch as well as the negative correlations with a, ash, and insoluble dietary fiber are likely a reflection of the increasing degree of refinement brought upon by tempering. The lack of a significant correlation with protein likely is due to protein contents remaining relatively constant after tempering; thus, the increase in starch came at the expense of insoluble fiber.

Table 4. Solvent retention capacity (SRC) of tempered intermediate wheatgrass flour.

Lactic Acid SRC	Incubation Time/Temperature					
	4 h		8 h		24 h	
Target Moisture (%)	30 °C	45 °C	30 °C	45 °C	30 °C	45 °C
NT [1]	93.7 ± 1.3 bA	96.9 ± 1.7 bA	88.9 ± 0.0 B	88.1 ± 0.9 bB	92.8 ± 0.8 bA	94.0 ± 0.8 A
12	100.8 ± 1.1 aA*	107.2 ± 0.9 a*	95.5 ± 0.4 B	105.5 ± 0.9 a	97.8 ± 0.8 bB*	105.4 ± 0.7 *
14	103.2 ± 0.2 a	102.8 ± 1.4 a	106.5 ± 8.0	104.4 ± 1.3 a	112.2 ± 6.1 a	109.6 ± 7.5
Sodium carbonate SRC						
NT	111.0 ± 0.2 *	115.9 ± 1.2 aB*	112.6 ± 3.9 *	129.0 ± 3.9 aA*	121.8 ± 4.7 a*	132.7 ± 1.9 aA*
12	109.2 ± 1.5 *	113.6 ± 0.3 ab*	108.1 ± 2.1 *	110.9 ± 3.1 b*	105.2 ± 2.8 b	112.1 ± 0.6 b
14	108.4 ± 0.8	110.1 ± 1.5 b	111.5 ± 1.4	113.7 ± 0.2 b	110.5 ± 3.2 ab*	114.2 ± 1.4 b*
Sucrose SRC						
NT	131.1 ± 1.4 b	130.4 ± 4.5 b	123.3 ± 0.5 b	126.7 ± 3.7	142.5 ± 14.2	143.8 ± 17.6
12	149.3 ± 3.2 a	150.2 ± 0.1 a	124.9 ± 2.2 b	132.2 ± 8.0	128.8 ± 10.9	139.9 ± 3.4
14	147.0 ± 6.2 a	147.4 ± 1.3 a	139.6 ± 3.9 a	143.7 ± 0.9	141.3 ± 3.7	140.9 ± 3.2
Water SRC						
NT	89.1 ± 0.4	88.4 ± 1.6 b	91.6 ± 1.5 a	86.3 ± 0.4 b	94.5 ± 3.6 *	87.9 ± 1.0 b*
12	87.4 ± 1.5 *	91.9 ± 1.0 aA*	86.3 ± 0.8 b	86.8 ± 1.7 bB	86.0 ± 3.2	89.1 ± 0.0 bAB
14	89.6 ± 0.9 *	93.1 ± 0.0 aAB*	89.1 ± 0.9 ab*	91.4 ± 0.5 aB*	92.1 ± 5.1	94.0 ± 0.8 aA

[1] NT, nontempered control flours. Lowercase letters show differences according to target moisture, uppercase letters represent differences due to incubation time, and asterisks indicate differences due to incubation temperature.

Swelling of flour in sodium carbonate solution is the result of starch damage [24]. However, sodium carbonate SRC followed a similar, yet less pronounced, pattern as all parameters that increased with bran contamination. In particular, nontempered samples incubated at 45 °C had significantly higher sodium carbonate SRC values than samples incubated for the same time to either 12% or 14% target moisture (Table 4). These results align with reports of tempering softening kernels and facilitating the separation of endosperm from bran layers without rupturing starch granules [29]. However, they were not significantly correlated to the results of the starch damage assay, which measures starch digestion via amylase. In addition, microscopy of granules would be informative. Sodium carbonate SRC exhibited significant, but moderate, correlations with L ($r = -0.527$, $p < 0.05$), a ($r = 0.566$, $p < 0.05$), insoluble fiber ($r = 0.605$, $p < 0.01$), starch content ($r = -0.557$, $p < 0.05$), and PPO activity ($r = 0.564$, $p < 0.05$).

Sucrose SRC is mediated by arabinoxylan swelling, which is an important constituent of IWG dietary fiber [30]. These results showed correlations of similar, moderate magnitudes than sodium carbonate SRC, but relationships were inverted: sucrose SRC had a significant, positive correlation with b values ($r = 0.485$, $p < 0.05$) and starch damage ($r = 0.614$, $p < 0.01$) and negative correlations with a ($r = -0.487$, $p < 0.05$) and protein ($r = -0.598$, $p < 0.01$).

SRC in water is affected by all flour polymers [24] but was a less informative parameter for our sample set. Only correlations with protein ($r = -0.627$, $p < 0.01$) and starch damage ($r = 0.500$, $p < 0.05$) were significant.

Previously, refined but nontempered IWG was reported to have lactic acid SRC, sodium carbonate SRC, sucrose SRC, and water SRC of 78.7, 98.1, 107.2, and 73.3, respectively [17]. Thus, the additional refinement achieved via tempering resulted in increases in SRC values for all solvents (Table 4).

3.6. Principal Component Analysis

The characteristics of the flour samples were summarized via a principal component analysis where the first two principal components accounted for >80% of the variability in the data set (Figure 3). Nontempered controls were distinctly separated from samples of 12% and 14% target moisture via principal component 1, which was negatively correlated with *a*, insoluble dietary fiber, ash, and PPO activity, while *L*, *b*, starch contents, and lactc acid SRC had high, positive correlations. Principal component 2 was positively correlated with protein contents and negatively correlated with water SRC. However, potentially because of the low variability in these properties, sample groups could not be effectively separated via principal component 2. As a result, differentiation between samples of 12% to those of 14% target moisture could not be achieved. In addition, samples could also not be separated based on incubation temperature or time (Figure S1a,b).

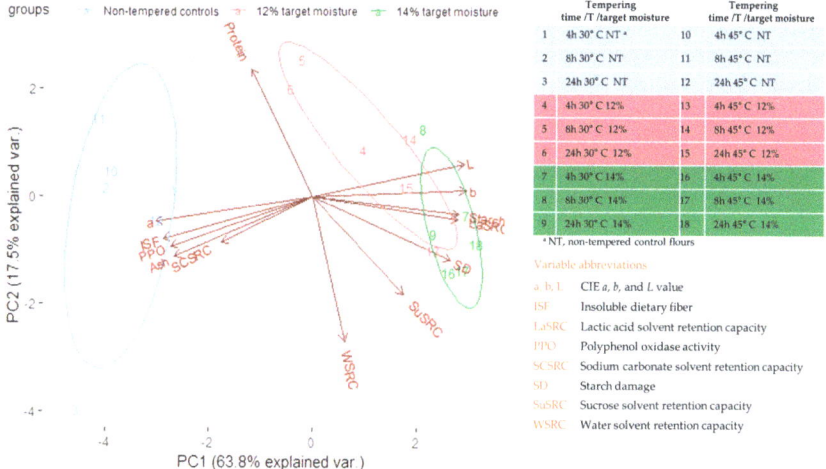

Figure 3. Principal component (PC) analysis of intermediate wheatgrass flours obtained via various tempering conditions.

4. Conclusions

This study has shown that tempered IWG samples have significantly different flour properties compared to nontempered controls. Rapid moisture uptake resulted in few significant differences in kernel or flour moisture among samples, which likely affected all other evaluated parameters. Tempering produced lighter, less brown, more yellow flour with more starch as well as more damaged starch, but it reduced polyphenol oxidase activity. To better establish the tempering parameters for IWG, future work will assess the functional attributes of tempered and refined IWG flours.

Supplementary Materials: The following are available online at http://www.mdpi.com/2304-8158/8/8/337/s1, Figure S1: Principal component (PC) analysis of intermediate wheatgrass flour obtained after different tempering treatments. Ellipses represent sample categorization via (a) temperature; (b) incubation time, Table S1: Pearson correlation coefficients among intermediate wheatgrass flour parameters.

Author Contributions: Conceptualization, G.A.A. and T.S.; methodology, C.T., G.A.A. and T.S.; formal analysis, C.T. and R.B.; investigation, C.T. and R.B.; resources, G.A.A; original draft preparation, C.T.; writing—review and editing, G.A.A., R.B. and T. S.; visualization, C.T.; supervision, G.A.A.; project administration, G.A.A.; funding acquisition, G.A.A. and T.S.

Funding: This research was funded through the Forevergreen Initiative at the University of Minnesota.

Acknowledgments: The authors would like to thank Sophie Held, Zihan Deng, and Leilany Vázquez Portalatín for help with sample preparation and analysis.

Conflicts of Interest: The authors declare no conflict of interest.

References

1. Sacchi, G.; Cei, L.; Stefani, G.; Lombardi, G.V.; Rocchi, B.; Belletti, G.; Padel, S.; Sellars, A.; Gagliardi, E.; Nocella, G.; et al. A Multi-Actor Literature Review on Alternative and Sustainable Food Systems for the Promotion of Cereal Biodiversity. *Agriculture* **2018**, *8*, 173. [CrossRef]
2. Kantar, M.B.; Tyl, C.E.; Dorn, K.M.; Zhang, X.F.; Jungers, J.M.; Kaser, J.M.; Schendel, R.R.; Eckberg, J.O.; Runck, B.C.; Bunzel, M.; et al. Perennial Grain and Oilseed Crops. *Annu. Rev. Plant Biol.* **2016**, *67*, 703–729. [CrossRef] [PubMed]
3. Cox, T.S.; Glover, J.D.; Van Tassel, D.L.; Cox, C.M.; DeHaan, L.R. Prospects for developing perennial-grain crops. *Bioscience* **2006**, *56*, 649–659. [CrossRef]
4. Jungers, J.M.; DeHaan, L.H.; Mulla, D.J.; Sheaffer, C.C.; Wyse, D.L. Reduced nitrate leaching in a perennial grain crop compared to maize in the Upper Midwest, USA. *Agric. Ecosyst. Environ.* **2019**, *272*, 63–73. [CrossRef]
5. Marti, A.; Qiu, X.X.; Schoenfuss, T.C.; Seetharaman, K. Characteristics of Perennial Wheatgrass (*Thinopyrum intermedium*) and Refined Wheat Flour Blends: Impact on Rheological Properties. *Cereal Chem.* **2015**, *92*, 434–440. [CrossRef]
6. Rahardjo, C.P.; Gajadeera, C.S.; Simsek, S.; Annor, G.; Schoenfuss, T.; Marti, A.; Ismail, B.P. Chemical characterization, functionality and baking quality of intermediate wheatgrass (*Thinopyrum intermedium*). *J. Cereal Sci.* **2018**, *83*, 266–274. [CrossRef]
7. Tyl, C.; Ismail, B.P. Compositional evaluation of perennial wheatgrass (*Thinopyrum intermedium*) breeding populations. *J. Food Sci. Technol.* **2019**, *54*, 660–669. [CrossRef]
8. Banjade, J.D.; Gajadeera, C.; Tyl, C.E.; Ismail, B.P.; Schoenfuss, T.C. Evaluation of dough conditioners and bran refinement on functional properties of intermediate wheatgrass (*Thinopyrum intermedium*). *J. Cereal Sci.* **2019**, *86*, 26–32. [CrossRef]
9. Kweon, M.; Martin, R.; Souza, E. Effect of Tempering Conditions on Milling Performance and Flour Functionality. *Cereal Chem.* **2009**, *86*, 12–17. [CrossRef]
10. Banjade, J.D.; Tyl, C.E.; Schoenfuss, T.C. Effect of dough conditioners and refinement on intermediate wheatgrass (*Thinopyrum intermedium*) bread. *LWT* **2019**, *115*, 108442. [CrossRef]
11. AACC International. *AACC International Approved Methods*; AACC International: St. Paul, MN, USA, 1999.
12. Fuerst, E.P.; Anderson, J.V.; Morris, C.F. Delineating the role of polyphenol oxidase in the darkening of alkaline wheat noodles. *J. Agric. Food Chem.* **2006**, *54*, 2378–2384. [CrossRef]
13. Hourston, J.E.; Ignatz, M.; Reith, M.; Leubner-Metzger, G.; Steinbrecher, T. Biomechanical properties of wheat grains: The implications on milling. *J. R. Soc. Interface* **2017**, *14*, 20160828. [CrossRef]
14. Oliver, J.R.; Blakeney, A.B.; Allen, H.M. The color of flour streams as related to ash and pigment contents. *J. Cereal Sci.* **1993**, *17*, 169–182. [CrossRef]
15. Gomez, M.; Pardo, J.; Oliete, B.; Caballero, P.A. Effect of the milling process on quality characteristics of rye flour. *J. Sci. Food Agric.* **2009**, *89*, 470–476. [CrossRef]
16. Siah, S.; Quail, K.J. Factors affecting Asian wheat noodle color and time-dependent discoloration—A review. *Cereal Chem.* **2018**, *95*, 189–205. [CrossRef]
17. Held, S.; Tyl, C.E.; Annor, G.A. Effect of Radio Frequency Cold Plasma Treatment on Intermediate Wheatgrass (*Thinopyrum intermedium*) Flour and Dough Properties in Comparison to Hard and Soft Wheat (*Triticum aestivum* L.). *J. Food Qual.* **2019**, *2019*, 8. [CrossRef]
18. Gibson, T.S.; Alqalla, H.; McCleary, B.V. An improved enzymatic method for the measurement of starch damage in wheat-flour. *J. Cereal Sci.* **1992**, *15*, 15–27. [CrossRef]
19. Wang, N.; Hou, G.G.; Dubat, A. Effects of flour particle size on the quality attributes of reconstituted whole-wheat flour and Chinese southern-type steamed bread. *LWT-Food Sci. Technol.* **2017**, *82*, 147–153. [CrossRef]

20. Singh, N.; Gujral, H.S.; Katyal, M.; Sharma, B. Relationship of Mixolab characteristics with protein, pasting, dynamic and empirical rheological characteristics of flours from Indian wheat varieties with diverse grain hardness. *J. Food Sci. Tech. Mys* **2019**, *56*, 2679–2686. [CrossRef]
21. Hammed, A.M.; Ozsisli, B.; Ohm, J.B.; Simsek, S. Relationship Between Solvent Retention Capacity and Protein Molecular Weight Distribution, Quality Characteristics, and Breadmaking Functionality of Hard Red Spring Wheat Flour. *Cereal Chem.* **2015**, *92*, 466–474. [CrossRef]
22. Liu, C.; Li, L.M.; Hong, J.; Zheng, X.L.; Bian, K.; Sun, Y.; Zhang, J. Effect of mechanically damaged starch on wheat flour, noodle and steamed bread making quality. *J. Food Sci. Technol.* **2014**, *49*, 253–260. [CrossRef]
23. Ma, S.; Li, L.; Wang, X.X.; Zheng, X.L.; Bian, K.; Bao, Q.D. Effect of mechanically damaged starch from wheat flour on the quality of frozen dough and steamed bread. *Food Chem.* **2016**, *202*, 120–124. [CrossRef]
24. Kweon, M.; Slade, L.; Levine, H. Solvent Retention Capacity (SRC) Testing of Wheat Flour: Principles and Value in Predicting Flour Functionality in Different Wheat-Based Food Processes and in Wheat Breeding-A Review. *Cereal Chem.* **2011**, *88*, 537–552. [CrossRef]
25. Belton, P.S. On the elasticity of wheat gluten. *J. Cereal Sci.* **1999**, *29*, 103–107. [CrossRef]
26. Duyvejonck, A.; Lagrain, B.; Pareyt, B.; Courtin, C.M.; Delcour, J.A. Relative contribution of wheat flour constituents to solvent retention capacity profiles of European wheats. *J. Cereal Sci.* **2011**, *53*, 312–318. [CrossRef]
27. Niu, Q.; Yu, P.; Li, X.; Ma, Z.; Hu, X. Solvent retention capacities of oat flour. *Int. J. Mol. Sci.* **2017**, *3*, 590. [CrossRef]
28. Lindgren, A.L. Solvent Retention Capacity and Quality Parameters of Whole Wheat Flour from Hard Red Spring Wheat. Ph.D. Thesis, North Dakota State University, Fargo, ND, USA, 2016.
29. Murray, J.C.; Kiszonas, A.M.; Wilson, J.; Morris, C.F. Effect of Soft Kernel Texture on the Milling Properties of Soft Durum Wheat. *Cereal Chem.* **2016**, *93*, 513–517. [CrossRef]
30. Schendel, R.R.; Becker, A.; Tyl, C.E.; Bunzel, M. Isolation and characterization of feruloylated arabinoxylan oligosaccharides from the perennial cereal grain intermediate wheat grass (*Thinopyrum intermedium*). *Carbohydr. Res.* **2015**, *407*, 16–25. [CrossRef]

© 2019 by the authors. Licensee MDPI, Basel, Switzerland. This article is an open access article distributed under the terms and conditions of the Creative Commons Attribution (CC BY) license (http://creativecommons.org/licenses/by/4.0/).

MDPI
St. Alban-Anlage 66
4052 Basel
Switzerland
Tel. +41 61 683 77 34
Fax +41 61 302 89 18
www.mdpi.com

Foods Editorial Office
E-mail: foods@mdpi.com
www.mdpi.com/journal/foods

www.ingramcontent.com/pod-product-compliance
Lightning Source LLC
LaVergne TN
LVHW071953080526
838202LV00064B/6737